Albert E. Seaton

The Screw Propeller

And Other Competing Instruments for Marine Propulsion

Albert E. Seaton

The Screw Propeller

And Other Competing Instruments for Marine Propulsion

ISBN/EAN: 9783954272730
Erscheinungsjahr: 2013
Erscheinungsort: Bremen, Deutschland

© maritimepress in Europäischer Hochschulverlag GmbH & Co. KG, Fahrenheitstr. 1, 28359 Bremen. Alle Rechte beim Verlag und bei den jeweiligen Lizenzgebern.

www.maritimepress.de | office@maritimepress.de

Bei diesem Titel handelt es sich um den Nachdruck eines historischen, lange vergriffenen Buches. Da elektronische Druckvorlagen für diese Titel nicht existieren, musste auf alte Vorlagen zurückgegriffen werden. Hieraus zwangsläufig resultierende Qualitätsverluste bitten wir zu entschuldigen.

THE SCREW PROPELLER:

AND OTHER COMPETING INSTRUMENTS FOR MARINE PROPULSION.

BY

A. E. SEATON, M.Inst.C.E., M.I.Mech.E., M.Council N.A.,

FORMERLY LECTURER ON MARINE ENGINEERING TO THE ROYAL NAVAL
COLLEGE, GREENWICH;
AUTHOR OF "A MANUAL OF MARINE ENGINEERING," ETC.

With Frontispiece, 6 Plates, 65 other Illustrations,
and 60 Tables.

LONDON:
CHARLES GRIFFIN & COMPANY, LIMITED.
PHILADELPHIA: J. B. LIPPINCOTT COMPANY.
1909.

PREFACE.

SOME thirty-two years ago, when engaged in putting forth the *Manual of Marine Engineering*, I could not find a single book on the Screw Propeller, or any text book containing such information on it as would enable a draughtsman to get out the leading dimensions, and much less to make a complete design of a screw suitable for any particular ship and conditions.

At that time John Bourne's admirable book on the Screw Propeller was not only out of print, but out of date, and I fear no engineer could at any time have designed a screw which would give satisfactory results from what was contained in it. It was, however, many years after that I first saw his work, and my wonder then, as now, is that such a book was allowed to disappear, seeing how much of interest it contained.

But perhaps the strongest comment on the knowledge of the screw propeller at that time is in the admission of the then Engineer-in-Chief of the British Navy, that he had settled the design of those for so very important a ship as the "Iris," intended to be the fastest ship in the Navy, by copying that of H.M.S. "Himalaya," a ship built in 1854, with the result that the British Admiralty, with all its knowledge and opportunities, with all its records of tests and trials, perpetrated a blunder never equalled in the history of steam navigation, although in the mercantile marine there had been not a few mistakes.

But even in modern times, notwithstanding the better knowledge and the aid of tank experiments, our best men do sometimes fail to achieve success, but the magnitude of the failure is inconsiderable compared with that of the "Iris."

In the *Manual of Marine Engineering* I attempted to supply the wants of designers by giving a rule or formula for each important dimension, generally based on scientific reasons, and always capable of giving results agreeable with the best and most successful practice.

Moreover, such rules were generally cast in such a form as to be quickly and easily used.

From time to time I have added to them and modified such of them as further knowledge had shown to require it, so that they have become generally applicable to the design of a screw for an Atlantic liner or a torpedo boat.

The object of the present work is to amplify and extend what was there mostly in skeleton or in rudimentary form; to add to it much that is of interest, although old; and to give all that is new and of importance and necessary to be known by the students, draughtsmen, sea-going engineers, designers, and others who have found the *Manual* of assistance to them.

The more abstruse and highly mathematical investigations connected with the theory of the resistance of ships and propellers have been left to be studied in the text-books of the schools and in the valuable papers contributed by Prof. Rankine, Prof. Cotterill, Mr R. E. Froude, Dr Froude, Prof. Greenhill, and others, to the Transactions of our professional institutions and learned societies.

Chapters I. and II. are devoted to the history of propellers in such a way as to give to the reader all the good inventions, their inventors, patent numbers, etc., and at the same time to revive and perpetuate the name of many a good pioneer among engineers. The young engineer of to-day is thus enabled to see what has been done by those gone before him, and to give the credit and praise to the right man.

In connection with this it is curious to note how often the important invention has emanated from someone outside the profession of engineering. For example, Francis P. Smith is described as a "farmer," Bennet Woodcroft as a "printer," Goldsworthy Gurney a "physician." The only "engineer" to whom it may be accorded that he did invent and patent a screw propeller as well as invent and make a locomotive engine long before anyone else, is the Cornishman Trevithick. But it is likewise a melancholy coincidence that except Gurney the above-named made no money by these inventions, and died poor men. It is true that poor Woodcroft had a crust given him by the Government in the form of an indifferently paid post in the Patent Office.

I am indebted to Mr C. De Graves Sells of Genoa for the particulars of the various groups of most interesting experiments made by his father, as also for other important information respecting the early history of the screw, and I take this opportunity of thanking him for these and some other acts of kindness I have experienced at his hands of a similar nature.

I also desire to tender my thanks to several other friends who have been kind enough to furnish me with information such as I was lacking, and which was necessary to my purpose, especially when engaged in determining what was really good modern practice.

In conclusion, I hope this book will have accorded to it the same kindly consideration as was experienced with the *Manual*, and that the readers will regard with favour the attempts to elucidate difficulties and to make known discoveries, rather than to scan too closely the many shortcomings it possesses, in fact, in the words of Prior—

" Be to her virtues ever kind,
And to her faults a little blind."

A. E. SEATON.

WESTMINSTER, S.W.,
March 1909.

CONTENTS.

CHAP.		PAGE
I.	Early History of Marine Propellers.	1
II.	Modern History of Propellers.	13
III.	Resistance of Ships	38
IV.	On Slip—Real, Apparent, Positive and Negative. Cavitation. Racing	54
V.	Paddle Wheels	69
VI.	Dimensions of Paddle Wheels	80
VII.	Hydraulic Propulsion: Internal Propellers and Jet Propellers	91
VIII.	The Screw Propeller: Leading Features and Characteristics; Thrust and Efficiency.	104
IX.	Various Forms of Screw Propeller	130
X.	The Number and Position of Screws	140
XI.	Screw Propeller Blades: their Number, Shape, and Proportions	153
XII.	Details of Screw Propellers, and their Dimensions	169
XIII.	Geometry of the Screw	186
XIV.	Materials used in the Construction of the Screw Propeller	194
XV.	Trials of the S.S. "Archimedes" and H.M.S. "Rattler".	202
XVI.	Trials of H.M.S. "Dwarf" and other Ships, made from time to time by the Admiralty	208
XVII.	Analysis of Mr Sell's Experiments made in 1856 with Propellers Six Inches, varying in Pitch and Surface Ratio	219
XVIII.	Experiments made by Isherwood and others.	226
	Index	248

ERRATUM.

Fig. 6, page 22, *for* " 1891 " *read* " 1841."

LIST OF ILLUSTRATIONS.

Frontispiece, Stern of the s.s. "Lusitania."

NO.		PAGE
1. Savery's Engine applied to Ship Propulsion		4
2. Paddle Steamer "Charlotte Dundas," 1802		9
3. Machinery and Wheels of Paddle Steamer "Comet," 1812		11
Plate I. Various Propellers	face page	14
,, II. ,, ,,	,, ,,	18
,, III. ,, ,,	,, ,,	20
,, IV. ,, ,,	,, ,,	22
,, V. ,, ,,	,, ,,	24
4. Francis P. Smith's Screw as first tried		18
5. Ericsson's Double Screws		19
6. Napier's Double Screws, 1841		22
7. Stern of H.M.S. "Rattler," 1843		23
8. Griffiths Early Patent Screw (self-adjusting)		25
9. Woodcroft's Adjustable Blades		26
10. Roberts' Patent Boss, 1851		27
11. H. Hirsch's Screw Propeller of 30 degrees		30
12. Thornycroft's Stern for Shallow Draught Screw Ships		35
12A. Yarrow's Drop Flap for Shallow Draught Screw Ships		37
12B. Kirk's Block Model		42
13. Influence of Screw on Water beyond Tip of Blades		55
14. Flow of Water to Paddle Float		56
15. Passage of a Curved Screw Blade Section through Water		62
16. Screw working with Negative Apparent Slip		62
17. Flow of Water from Stern to Screw		63
Plate VI. Feathering Paddle Wheel in Motion	face page	72
18. Paddle Wheel of Dublin R.M.S., Inside Feathering Gear		73
19. Paddle Wheel of "Normandy" (no outer rim)		75
20. Stern Wheel Steamer, Single Wheel		77
21. Stern Wheel of the "Endeavour": Pair of Wheels, Feathering Floats		78
22. Details of Feathering Gear of a Paddle Wheel		87
23. Details of Float Bearing, etc., of a Feathering Wheel		88
24. Locus of Float Centres		89

LIST OF ILLUSTRATIONS.

NO.		PAGE
25.	Ruthven's Hydraulic Propeller, as in H.M.S. "Waterwitch".	94
26.	Thornycroft's Hydraulic Motor in a Torpedo Boat	
27.	Bessemer's Hydraulic Propeller	7
28.	Screw Blade on a True Helix	104
29.	Increasing Pitch Helix (Woodcroft's Wheel)	106
30.	Screw Blade bent forward, Griffiths' Patent	106
31.	Screw Blade thrown back by making central line Spiral on the Bed	107
32.	Screw Blade thrown back by Coning the Bed	107
33.	Screw Blade Curved Back	108
34.	Curves of Friction, Resistance, etc., of a Common Screw Blade	111
35.	Curves of Friction, Resistance, etc., of Griffiths' Screw	112
36.	Modern High Revolution Screw compared with that of H.M.S. "Rattler," 1845	113
37.	Froude's Curve of Indicated Thrust	122
38.	Propeller, on Griffiths' Patent (1860), Adjustable Blades	131
39.	Mangin's Double Screw	132
40.	Hirsch's Screw of 60 degrees	135
41.	Mercantile Four-bladed Propeller	136
42.	Oval Blade of Equal Surface for Different Diameters	137
43.	Modern Bronze Naval Screw	138
44.	Screw outside of Rudder	148
45.	Auxiliary Screw outside the Rudder	149
46.	Phipps' Lowering Screw, 1850	150
47.	Bevis' Feathering Screw	160
48.	Maudslay's Feathering Screw and Banjo Frame	161
49.	Curve of Bending Moments	174
50.	Typical Screw Blade Section	174
51.	Solid Cast Iron Propeller Blade	178
52.	Various Root Sections of a Screw Blade	179
53.	Longitudinal Section of Typical Screws of Equal Area of Blade	179
54.	Method of Delineating a True Screw Accurately	187
55.	Simple Method of Delineating a Screw	189
56.	Increasing Pitch Blade	190
57.	Pitch Measuring Instrument	192
58.	Bent Propeller Blade of Parson's Manganese Bronze after Stranding	196
59.	Bent Propeller Blade of Stone's Bronze	197
60.	Bent Propeller Blade, Manganese Bronze Co.	200
61.	Blades dovetailed into Forged Boss	201
62.	Screws tried on H.M.S. "Iris"	216
63.	Screws used by Isherwood in his Experiments	226

MARINE PROPELLERS.

CHAPTER I.

EARLY HISTORY OF MARINE PROPELLERS.

It is more than probable that the first instrument whereby motion was imparted to a vessel floating in deep water was the human hand used as a paddle. It would soon dawn on the mind of the intelligent navigator of such early days that there was far too little resistance on the part of the hand, either with the fingers side by side, or extended to get the best results from the muscular efforts of the arms; and although no word would be found for the phenomenon, we should say now there was too much slip from want of sufficient acting surface. To remedy this defect, art would supplement Nature, and its first step would be to fit to the hand large shells or pieces of flat wood. No doubt the next step of art would be the substitution of a shank or handle for the forearm, which had no doubt suffered from continuous immersion in cold water; in this way a hand paddle would be evolved, which answered the purpose of those early mariners just as it does now for the modern barbarous nations on the rivers and lakes in various parts of the world, whose boats so propelled run side by side with the steam launches and steam stern-wheelers of the white man.

All such paddles so impelled impart motion to some of the water with respect to the still water through which the boat has to pass, and the reaction from this disturbed water is taken at the shoulders of the man and imparted through the body to the boat so as to produce the motion of it in the opposite direction to that taken by the disturbed water.

In course of time art advanced the propeller another step by lengthening the handle so as to permit of it resting on the gunwale or edge of the boat, and of acting against a peg or notch on it, thus permitting of the muscular effort of the rower to be extended beyond that of his arms; and although the reaction from his body is then in the opposite direction from that in which the boat is intended to travel, the action at the gunwale is in excess of this, due to the leverage of the paddle or oar, so that the resultant pressure produces motion in the right direction.

Propellers on this principle (viz. the paddle or oar) of various forms and dimensions are now used throughout the civilised world for all kinds of boats and barges, and in the past for many centuries were employed, especially in the Mediterranean, for the propulsion of all ships, even those of very large size, when there were two, or sometimes three, rows of rowers on each side.

How, or when, or by whom it was discovered that the motion could be induced by the oscillating of a single oar placed in a notch at the stern of the boat so that the blade moved obliquely to the water stream somewhat as a blade of a screw propeller does, is not known. It is certainly a very convenient form of propulsion for a single man to adopt, and may possibly have led more than one mind to the idea of the screw propeller, in much the same way as the paddle wheel was no doubt evolved from the idea of getting continuous instead of reciprocating motion with paddles by mechanical means.

All that is necessary for a book of this kind in tracing the early development of propellers is to show that the principle involved is in every case the same, and that, while improvements have gone on and do go on from time to time, there is no departure from the principle that the motion of every self-propelled ship is due to the projection by its propeller of a stream of water in a direction opposite to that in which it is intended the ship shall move. The only exception that can be cited is that of the ferry boats and river craft which are moved by means of a chain or rope submerged and operated upon by a winch arrangement in the ship itself.

There is every reason to believe that paddle wheels were used as a means of propulsion in quite early times, even as far back as the time of the Consul App. Claudius, who obtained the cognomen of "Caudex" because he employed boats propelled by paddle wheels to transport his troops into Sicily (B.C. 264).

The Chinese as well as the ancient Egyptians are more than suspected of having a knowledge of the use of a paddle wheel for ship propulsion.

In A.D. 1472 there is the evidence of R. Valturius that paddle wheels were in use instead of oars, inasmuch as he shows a view of two galleys having wheels (five pairs) on each side, each pair running on one common axle with a crank in the middle and the cranks connected together.

In A.D. 1543 it is stated that one Blasco de Garay, a Spaniard resident at Barcelona, propelled a vessel by an engine " consisting of a large caldron or vessel of boiling water and a moveable wheel attached to each side of the ship." This is the first record of a steam ship, and it is possible that the engine may have been on Hero's principle, and therefore of the nature of a turbine.

In A.D. 1578 it is related by one W. Bourne that "you may make a boat to go without oars or sail by the placing of certain wheels on the outside of the boats in that sort that the arms of the wheels may go into the water and so turning the wheels by some provision, the wheels shall make the boat go."

In A.D. 1597 Roger Bacon, writing, said that "We have seen and used in London, a warlike machine driven by internal machinery, either on land or water." "Succeeding years have shown us a vessel which being almost wholly submerged would run through the water against waves and wind, with a speed greater than that attained by the fastest London pinnaces."

In A.D. 1663 the Marquis of Worcester claims, among other matters, to have invented a boat with paddle wheels on an axle across it, which axle is turned by the action of the stream on the paddles.

In A.D. 1681 the celebrated philosopher Robert Hook describes certain windmills which, as stated by Mr Bourne, "had all the main features of the screw propeller and feathering wheel."

In A.D. 1682 it is related that a "tow vessel" was used at Chatham Dockyard for moving the ships in the river by means of paddle wheels on an axle turned in some way by a horse, probably by means of a gin.

In A.D. 1690 D. Papin described his steam cylinder containing a piston which was forced downwards by the pressure of the atmosphere when the steam below it had been condensed, and stated that it might be applied, among other things, to the propulsion of ships by paddle wheels, such as he had seen made in London for Prince Rupert, to be turned by horses. He proposed to use two or three of his cylinders with a rack arrangement.

In A.D. 1693 M. Du Quet tried as a propeller four oar-shaped blades set in a wheel so that they could be moved through 90 degrees, and so that each one after propelling could be made to feather and return edgeways.

In A.D. 1698 Thomas Savery, in his patent No. 356, relating to the raising of water and occasioning motion to all kinds of mills by the impelling force of the steam engine, states: "I believe it may be very useful to ships, but I dare not meddle with that matter, and leave it to the judgment of those who are the best judges of maritime affairs." And again: "As for fixing the engines in ships when they may be thought probably useful, I question not but we may find conveniences for fixing them."

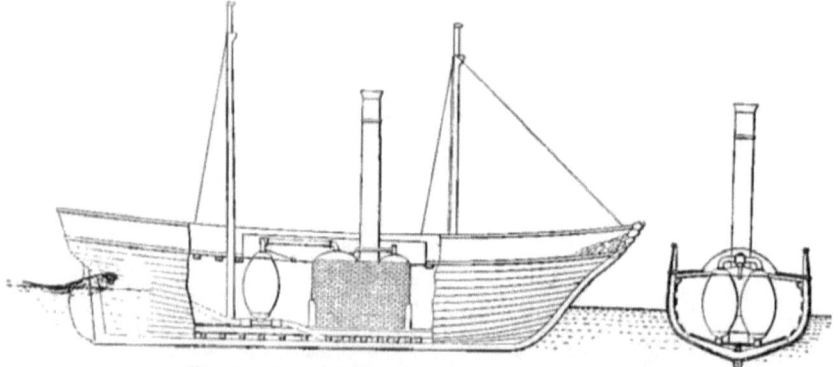

FIG. 1.—Savery's Engine applied to Ship Propulsion.

In A.D. 1707 it is stated in letters addressed to Leibnitz that Dennis Papin had used one of Savery's engines to propel a ship or boat on the river Fulda.

In A.D. 1721 experiments were made in France by M. Du Quet with paddle wheels of a sort, as much as 18 feet in diameter, three of which were fitted on each side of a galley and operated by as many as two hundred men.

In A.D. 1729 John Allen took out a patent claiming, among other things, to navigate a ship in a calm. "My method will be effected by forcing water or some other fluid through the stern or hinder part of a ship, at a convenient distance under the surface of the water, into the sea, by proper engines placed within the ship. Amongst the several and peculiar engines I have invented for this purpose is one of a very extraordinary nature, whose operation is owing to the explosion of gunpowder, I having found out a method of firing gunpowder *in clauso*, or in a confined place, whereby I can apply the whole force of it, which is inconceivably great, so as to communicate motion to a great variety of engines; which may also be applied for the draining of mines and other purposes."

This is most interesting as being the first instance of a proposal to propel a ship by jet or stream of water, and further, that the power was to be by means of what is known as an internal combustion engine.

In the same year, A.D. 1729, Du Quet, who seems to have been very interested in marine propulsion, called attention to a method for towing vessels by submerging a screw or helical frame having eight vanes fixed to the end of as many spokes and inclined at an angle of 54 degrees. The screw was caused to revolve by the stream of water, and by gearing to impart motion to a windlass around which the tow rope was wound.

In A.D. 1730 Mr Allen makes a further suggestion that application of the fire-engine should be made to work pistons for propelling a vessel by forcing out air or water under the surface of the internal water.

In A.D. 1734 Duviver gives a description of a ship having paddles fixed to a frame, very much like that which is now known as the lazy-tongs, on each side and worked by the oscillating of a heavy pendulum.

In A.D. 1736 Jonathan Hulls took out a patent which is interesting, inasmuch as it is claimed for him by his descendants that he was the first to produce an actual steam-propelled ship. His own claim was for a machine "for carrying ships and vessels out of or into a river or harbour against wind and tide in a calm." In his practice, it appears from such records as there are that his vessel was what is now known as a "stern-wheeler," the paddle wheel being carried in a framework at the stern of the vessel. The circular motion of the wheel was given by means of ratchet wheels operated on by ropes, one of which was connected with the piston of a condensing engine and the other to a falling weight, the weight being raised by the excess of the power of the steam piston.

In A.D. 1738 D. Bernouilli proposed to propel ships by forcing water through orifices so as to make jets to flow towards the stern, somewhat as Allen had suggested in 1729, and as shown in fig. 1.

In A.D. 1752 D. Bernouilli described a method of propelling vessels by wheels with vanes "set at an angle of sixty degrees, both with the arbor and keel of the vessel, to which the arbor is placed parallel. To sustain this arbor and the wheels, two strong bars of iron, of between two and three inches thick, proceed from the side of the vessel at right angles to it, about two feet and a half below the surface of the water." From this it would appear that practically a submerged screw propeller was intended. It is also noticeable in A.D. 1746, Bouger speaks of revolving arms like the vanes of a windmill as having been tried for the propulsion of ships.

In A.D. 1760 J. A. Genevois proposed to employ a propeller having hinged blades so as to open and shut like the back of a book, which would be full open on making the effective stroke, and closed again on the return so as to offer little obstruction. No doubt this idea was suggested by observing the foot of a duck or other water bird. This was followed years afterwards by a whole series of patents, none of which were successful, and all proved a source of extreme disappointment to their promoters. It is interesting to note in passing that this man was the first to suggest the use of water distilled from the sea for drinking and other purposes on board ship.

In A.D. 1770 James Watt made a suggestion for a screw propeller turned by one of his engines, but he did nothing further; in fact, he was much opposed to the use of his steam engines on board a ship.

In A.D. 1776 John Barber claimed, among other things, in his patent No. 1118, to apply the turbine on Branca's type to the propelling of ships, and provided for reversing the motion by making the nozzles reversible.

In A.D. 1778 a Jesuit missionary described a paddle-wheel boat he had seen at Pekin, China, as being 42 feet long by 13 feet broad, propelled by a paddle wheel on each side having flat arms whose ends dipped in the water to the extent of a foot; it was moved round by men.

In A.D. 1780 James Pickard of Birmingham took out his patent for a crank as a means of transforming the lineal motion of a steam engine piston into circular motion as in the turning of wheels.

In the same year Jouffroy used an engine with two cylinders to propel a boat having duck-foot-shaped propellers; a year later he tried a steamboat on the Rhône of a similar kind, but it was worked with paddle wheels.

In A.D. 1783 Jouffroy is said to have built a steamship 140 feet long, having a pair of wheels on one shaft moved by a single cylinder engine.

In A.D. 1785 Joseph Bramah, in his patent No. 1478, claimed to fit a wheel with inclined fans or wings, similar to the fly of a smoke jack or the vertical sails of a windmill; from this and other descriptions in the patent there is no doubt that Bramah had the complete idea of the screw propeller working at the end of a shaft projecting from the stern of a vessel and wholly submerged. (*Vide* fig. 1, Plate I.)

In A.D. 1786 John Fitch of Philadelphia, U.S.A., had built by Messrs Brookes & Wilson a boat 45 feet long and 12 feet beam, fitted with a steam engine having a cylinder 12 inches in diameter and 3 feet stroke, running at forty revolutions, operating on twelve oars or paddles so arranged that half of them were acting when the other half were out of action. It was tried in August 1788, but failed to make a higher speed than three miles per hour.

In A.D. 1787 Fitch had another boat built, 60 feet long, 8 feet beam and 4 feet deep, and fitted with the machinery of the above described boat, but a better system of paddles was fitted, they being so arranged as to have the same motion imparted to each of them as to a paddle worked by hand. The speed of this ship was somewhat better than the former, but still low.

In A.D. 1787 William Symington patented a means of obtaining rotary motion. In the same year Patrick Miller of Dalswinton, N.B., published his pamphlet on propelling ships by paddle wheels turned by men; it gave also the results of his experiments.

In A.D. 1788 Robert Fourness and James Ashworth took out their patent for raising the paddle wheels out of the water when not required.

A.D. 1788 is chiefly famous, however, for the trial of the first practicable steamship of which we have a definite record and accurate particulars. William Symington in that year was introduced to Patrick Miller's notice by Mr James Taylor, and his steam engine, which could be worked without an air pump, was chosen as suitable for the purpose of propelling Mr Miller's double-hulled pleasure boat, 25 feet long by 7 feet beam, which he had built at Dalswinton the preceding winter, by means of a paddle wheel working between the hulls. The engine was not much more than a toy, as the diameter of its two cylinders was only 4 inches, and it was mounted on a portable frame on the deck. It was, however, sufficiently powerful to move the ship at the rate of five miles an hour. In this pioneer work honour is due to Symington, the engineer who designed and superintended the construction of the machinery; to Patrick Miller, who found the capital and built the ship; and lastly, to James Taylor, who brought them together and acted the part of "amicus curiæ" throughout. Strange to say, he is the only one who had a reward for this valuable experiment, for his wife enjoyed a Government pension of £50 per annum. Patrick Miller having spent a fortune on it and other experiments, let the question of marine propulsion drop.

In the same year James Rumsey proposed to propel ships by drawing water through an orifice near the bow into an ordinary vertical pump and expelling it by a tube through the stern; but his ideas on propulsion were vague as well as numerous. He succeeded, however, in getting a wealthy American residing in London to pay for the construction of such a boat, and in 1793 she was tried on the Thames and attained a speed of only four knots, for although the pump was 24 inches diameter, the discharge orifice was no more than 6 inches square.

In A.D. 1789 Fitch had another boat built, named the "Thornton," fitted with more powerful machinery than supplied to the previous ones, so that the speed attained by it was at the rate of eight miles an hour.

In A.D. 1790 Fitch applied for a patent for forcing air and water by means of steam, through trunks, which was therefore a mode of hydraulic propulsion. It is also said that about 1791 he applied for a patent for paddles, both as side wheels and stern wheels. He appears to have been ingenious, persevering, and a good mechanic, but a dissolute man. In 1798 he committed suicide.

In A.D. 1792 James Rumsay took out another patent, No. 1903, in which he claims a centrifugal pump, etc.; he proposed to place a screw in a frame between the hulls of two boats connected together "whose axis being moved by horses, by steam, or by men, or any other power applied to the cog wheel, when the boat will take motion." The axle and screw may be used "before or behind a single boat or in the bottom."

In A.D. 1793 a paddle boat worked by steam was actually run by one John Smith on the Bridgewater Canal from Manchester to Runcorn. The engine was one of Newcomen type working with a beam, a connecting rod and crank; each paddle wheel had seven arms or floats. The speed was, however, only two miles an hour.

In A.D. 1794 William Lyttleton patented (No. 2000) an arrangement of three helical strips or threads projecting out of a cylinder hung in a frame and submerged either at the bow or stern of a ship. This propeller, which was of course a screw, was rotated by an endless rope from on deck. Here, again, when tried, the speed obtained by it was disappointing, being only two miles per hour. (See fig. 2, Plate I.)

In A.D. 1798 Robert Fulton stated he had, in this year, tried a four-bladed screw as a propeller on a boat.

In A.D. 1800 Edward Shorter patented, as a means of propelling ships, the fitting of a perpetual sculling machine to the stern of a ship, consisting of a two-bladed screw at the end of a revolving shaft set at an angle like an oar in sculling, and having a universal joint connection with a horizontal shaft on the deck, so that the screw could be raised and lowered to suit the trim of the ship. The screw shaft end was supported from a float and steadied in place by guy ropes.

In A.D. 1801 William Symington patented the fitting of a connecting rod from the piston rod end to the crank pin of a paddle

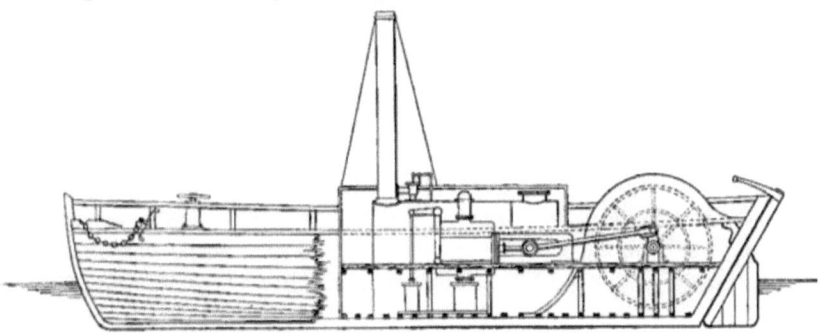

FIG. 2.—Paddle Steamer "Charlotte Dundas," 1802.

shaft, as seen always in stern-wheelers to-day. He applied this to the "Charlotte Dundas" in 1802.

In A.D. 1802 Shorter's screw arrangement was tried on board H.M.S. "Dragon" and "Superb," and to the transport "Doncaster," which latter ship attained a speed of 1½ miles per hour when deeply laden, with eight men only at the capstan working it.

In A.D. 1802 William Symington constructed the tow boat "Charlotte Dundas" for Lord Dundas of Kerse, N.B. This may be said to be the first steamship to be used for practical purposes as well as experiment; and but for the fears of damage to their canal by the proprietors of it, this ship might have been regularly employed, and the general use of other steamships would probably have followed at once. As it was, this little ship, after she had demonstrated her power by towing two barges, each of 70 tons burden, 19½ miles in six hours against a strong wind, was laid up and made no use of.

The "Charlotte Dundas" was what we now call a *stern-wheeler*, having one paddle wheel at the stern turned by a horizontal double-acting engine having one cylinder 22 inches in diameter and 4 feet stroke, directly connected by a rod as before described; in fact, quite like a present-day engine for the same purpose. She seems to have been about 44 feet long over all.

In A.D. 1804 John Stevens in America made and tried a steamboat having a screw propeller on Bramah's plan worked by a rotatory engine. The latter, however, was not a success, and was replaced by a Watt engine, when the speed attained was four miles an hour. The enterprise, however, proved a failure, owing to boiler troubles consequent on its being tubular and novel in design and too small for the engine.

In A.D. 1807 Robert Fulton of New York produced his famous steamer the "Clermont," of 160 tons burden, 130 feet long, 16½ feet beam, and 7 feet deep. She was propelled by a pair of side wheels 15 feet diameter, having floats 4 feet long dipping 2 feet into the water, operated by an engine made by Bolton & Watt of Birmingham, England, having one cylinder 24 inches diameter and 4 feet stroke supplied with steam by a boiler 20 feet long, 8 feet broad, and 7 feet high. On trial her speed over a run of 110 miles was at the rate of 4·6 miles per hour; the following day she did forty miles in five hours. The following year she was lengthened to 140 feet of keel, and then attained a mean speed of quite five miles per hour. Mr Fulton built other equally successful steamers, the largest being the "Paragon," of 331 tons burden and as much as 173 feet long. This was in 1811, when the "Comet" was being projected by H. Bell in Scotland. Very many years elapsed before a steamer of this length was built in Great Britain. Fulton, like Symington, Taylor, Bell, and so many other pioneer engineers, ended his days, in 1815, in penury.

In A.D. 1811 Henry Bell of Helensburgh, N.B., had built by John Wood & Co., Port-Glasgow, a little vessel of 30 tons burden, 40 feet long and 10½ feet beam, named the "Comet." He fitted her with a side lever engine having a single cylinder 11 inches diameter and 16 inches stroke, whose crankshaft was geared to two shafts, one before and one abaft the engine; each had a paddle wheel at each end, and was therefore what a locomotive engineer would call "a two-pair coupled" job. She attained a speed of five knots, and traded between Greenock and Glasgow; after being lengthened 20 feet and fitted with one pair of complete wheels and a new cylinder 12½ inches

EARLY HISTORY OF MARINE PROPELLERS. 11

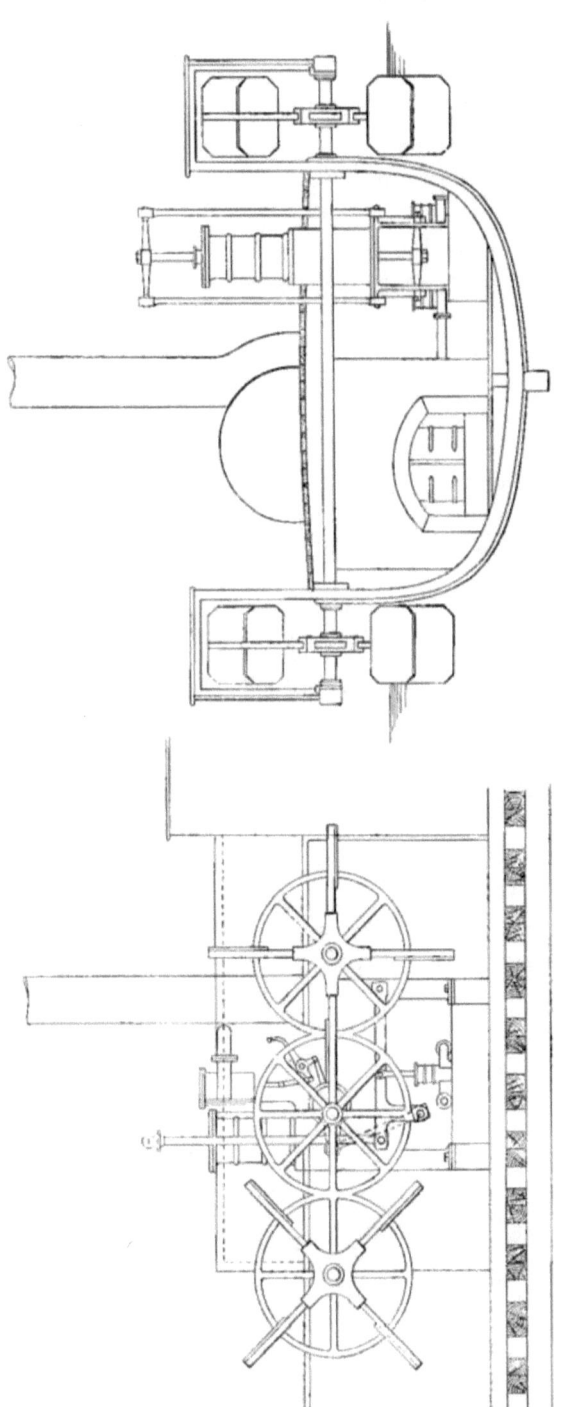

Fig. 3.—Machinery and Wheels of Paddle Steamer "Comet," 1812.

diameter her speed seems to have been 7·8 miles per hour or 6¾ knots. She was wrecked at Crinan, N.B., in 1820 in the tide race.

In A.D. 1815 the steamship "Thames" performed the voyage from Glasgow, where apparently she was built, to London, successfully. She called at Dublin on the road, and had Dr Dodd as a passenger.

The construction of steamers for commercial purposes now became general in various parts of the kingdom, even at those remote from the sea; for example, the paddle steamer "Britannia," of 50 tons, 65 feet long and 13 feet beam, was built at Gainsborough, Lincolnshire, in 1816, and sold to Portsmouth the following year. In 1817, at the same place, another builder produced the "Prince of Coburg," of 71 tons, 76·5 feet long, 14·4 feet beam, with engines by Aaron Manby of Staffordshire, and sent her to the Solent. Three years later Richard Pearson built at Thorne, near Doncaster, the paddle steamer "Kingston," 120½ tons, 106 feet long and 20 feet beam, followed by the "Yorkshireman" a year later, of 164¼ tons, 120 feet long, 21 feet beam. The latter had geared engines made by the Butterly Company, Derbyshire, wherein the engine shaft ran three times as many revolutions as the paddle shaft.

In A.D. 1822 Marc Isambard Brunel took out a patent for two inclined cylinder engines, the cranks at right angles, the piston rods fitted with roller guides, and the weight of the piston relieved by "spring supporters" on the extremities of the "head beam," the engine to be governed by means of a stream of water pumped through an orifice. The condenser was to be formed of an assemblage of pipes which collectively formed a spacious chamber. They were to be connected together with a set of smaller pipes, and the whole placed in an iron reservoir, thus forming a surface condenser.

It was not till 1822 that the Admiralty indulged in a steamship of their own, when Oliver Lang, the master shipwright of Deptford, built for the navy a small tug or tender, which, strange to say, their Lordships named the "Comet," as another mysterious visitor had crossed the heavens since the one from which Bell named his little pioneer ten years before. Ten years afterwards their Lordships ventured to build the "Salamander" paddle steamer, and fit her with guns as a warship.

CHAPTER II.

MODERN HISTORY OF PROPELLERS.

MANY years were to elapse from the time of the trials of Shorter's screw in H.M.S. "Dragon" before the screw was again put to a real practical test as a propeller; to the paddle wheel, having been proved successful for steam navigation both in smooth water and rough seas, men's minds and money were directed rather than to the out-of-sight screw, as is shown by the huge number of patents taken out for new and sometimes improved wheels or accessories.

The most important of them was the Morgan wheel, applied for by Elijah Galloway and sealed in July 1829 (No. 5805), which after running the full fourteen years, was renewed for a further five years, and consequently held good until 1848. The object of the inventor was to so manipulate the floats that at entry and emersion they were at the angle or inclination they would have occupied on a radial wheel of much larger diameter. This will be described in detail in another chapter; suffice it to say here that Hook had anticipated the *idea* so far back as 1683.

A.D. 1815.—Richard Trevithick, the ingenious Cornishman, suggested the screw as a fit instrument for marine propulsion in connection with his high-pressure engines, and the form he chose was that of leaves or blades placed obliquely on a cylindrical axle, and in some cases he would have it to revolve in a fixed cylinder; in others, the cylinder revolved with it; but generally the screw revolved without any cylinder surrounding it. "It may revolve at the head or the stern of the vessel; or one or more such worms may work on each side of the vessel," he said.

A.D. 1816.—John Millington of Hammersmith took out a patent for the use of a screw as a good method of propelling vessels having "two vanes, each extending to about a quadrant of a circle, as they produce a greater effect than any other number." He claims "to fit a propeller either at the head or stern of vessels, or at both of them at the same time." Millington's propeller shaft was set at an angle so that its inner end was above water, and attached to a horizontal shaft going inboard by means of a Hook's universal joint. The outer end was suspended from a spar on the ship's end, and arranged to be raised and lowered to suit the immersion. He also proposed to fit guy ropes on either side so that the spar might have motion horizontally, and so cause the propeller to steer the ship. (See fig. 3, Plate I.)

A.D. **1816.**—In this year William Church of Birmingham took out a patent for a propeller consisting of two wheels revolving in opposite directions like Perkins'. Church, however, proposed a number of bent paddles placed upon cylindrical rings, set so as to work in opposite directions, and they might be placed within a fixed cylinder. (See fig. 6, Plate I.)

A.D. **1816.**—Richard Wright took out his patent No. 4088, in which he claims, among other things, to make a feathering paddle wheel with a somewhat complicated apparatus, which he describes, and claims that it will cause the floats to enter the water vertically, which, of course, is not what is best with the ship when under way.

He also claimed to fit a two-cylinder two crank (at right angles) compound engine with a receiver between the cylinders.

A.D. **1817.**—Joseph Claude Niepce claims, in his patent No. 4179, to propel a ship by expelling water alternately from two reservoirs in a stream through the stern of the ship, and to use as his expelling force "expanded air produced by the combustible matter in a receiver which escapes through apertures closed with valves, etc., which will be like the force of steam by pressing upon the surface of the water enclosed in the receivers." In other words, an internal combustion apparatus used direct as a means of propulsion; he proposed using volatile oils for the purpose.

A.D. **1818.**—A regular line of steamers was established by Mr David Napier for service between Glasgow and Belfast, and Mr Dawson established a similar service of steamers on the Thames from London to Gravesend.

A.D. **1819.**—The first steamship, the "Savannah," 350 tons, fitted with lifting paddle wheels, crossed the Atlantic from New York to Liverpool, but the voyage was done mostly under sail; the engine being used only eighteen days out of the thirty-five taken in crossing; it consisted of one cylinder 40 inches diameter and 72 inches stroke.

PLATE I.

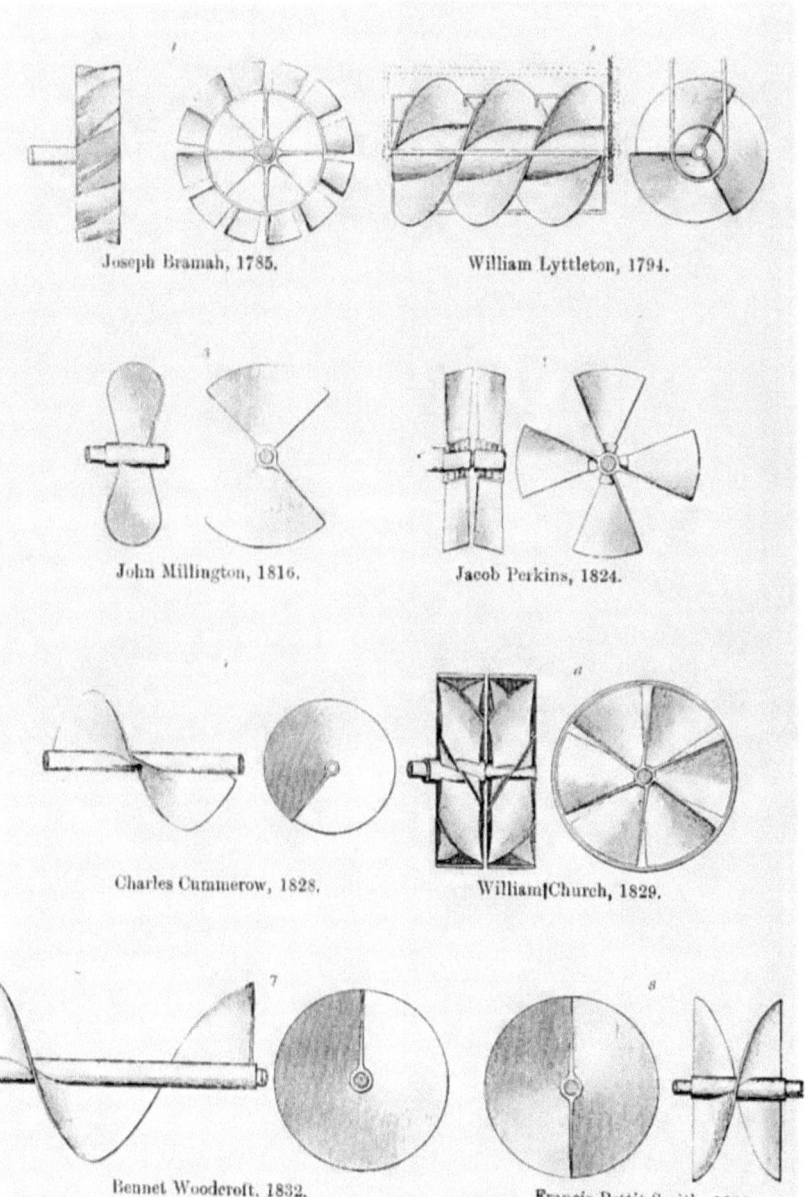

Joseph Bramah, 1785.

William Lyttleton, 1794.

John Millington, 1816.

Jacob Perkins, 1824.

Charles Cummerow, 1828.

William Church, 1829.

Bennet Woodcroft, 1832.

Francis Pettit Smith, 1836.

A.D. 1822.—H.M.S. "Comet," paddle steamship, built at Deptford for the Admiralty.

A.D. 1824.—Jacob Perkins patented No. 4998, an arrangement of screw propeller by which two blades were fixed on the end of a hollow shaft through which a second shaft passed, also having a pair of blades on its end, but set to the opposite "hand" of the former, so that when revolving in different directions they acted together in projecting a stream of water. (Fig. 4, Plate I.)

He is also the first to prescribe a varying pitch, as he proposed to make the blades with an inclination of 45 degrees at the boss and $22\frac{1}{2}$ degrees at the tips.

Since his day this idea of a right-handed and left-handed screw combination has been patented many times, notably by Ericsson in 1836, George Smith in 1838, and others until 1853, when a similar application from John Pym was refused protection by the Patent Office.

It is recorded that in this year John Swan tried double-bladed screws, one on each side of the ship, and fully immersed. This would appear to be the first attempt at twin screw propulsion.

A.D. 1825.—Samuel Brown patented (No. 5126) the idea of propelling such a ship as a ferry boat by a chain secured at the ends lying in the water and passing round a wheel on the ship, which is turned by machinery.

In the same year Samuel Brown gained a reward of a hundred guineas for the best suggestion for propelling ships without paddle wheels, by proposing a ship with a screw at the bow. Such a ship was built and tried on the Thames, attaining a speed of six to seven miles per hour; the screw, however, was rotated by a Brown "gas vacuum" engine, which gave trouble to such an extent as to discredit the whole undertaking and bring about the bankruptcy of the Company. This failure was, no doubt, also the means of postponing the adoption of the screw as a marine propeller for many years.

In A.D. 1825 also was witnessed the successful application of steam to ocean navigation, the steamer "Enterprise" having made the voyage from England to Calcutta in 113 days. This little ship was only 470 tons burden, 122 feet long, and 27 feet beam.

A.D. 1827.—William Hale patented No. 5594, an arrangement of screw having one or more threads turning on its axis in a vertical cylinder, drawing water through the bottom of a ship and expelling it through the stern to obtain motion. He says he found "two threads in the screw are better than one."

A.D. 1828.—Charles Cummerow, in his patent No. 5730, manifests a grasp of the practical problems involved in screw propulsion. He prescribed a screw of a single convolution with a pitch equal to half the diameter; he proposed to fit it in the deadwood of the ship and to support it at the outer end of the after-stern post on which the rudder is to hang; the shaft is to pass inboard through a tube and stuffing-box, and finally proposed to gear it to the engine shaft so that the screw should turn three times per second—that is, 180 revolutions per minute. (See fig. 5, Plate I.)

A.D. 1829.—Archibald Robertson patented, No. 5749, the idea of fixing the floats of a paddle wheel obliquely " at an angle of from 40 to 70 degrees to the plane of the wheel's motion " and parallel to each other. Also to place the paddle shafts obliquely.

Jacob Perkins in the same year proposed, in his patent No. 5806, to place the floats of a paddle wheel at an angle of 45 degrees to the plane motion, and the shafts at the same angle, so that when working the immersed floats would be perpendicular to the ship's keel.

In this year Elijah Galloway patented the well-known feathering wheel generally called the Morgan. His method of feathering remains in use to-day, while all others have disappeared. A full description of it is given in Chapter IV.

In the same year Julius Pumphrey claimed, in patent No. 5765, " to use two spirals at the stern, one on each side of the rudder, in the direction of the ship's length, wholly under water, so that the rudder may act freely between the shafts of the spirals; and cutting the shafts off opposite the stern posts or hangings of the rudder, to connect them again by Hook's universal joint, and carry the spirals with a strong frame which is hung to turn in concert with the rudder. The spirals are to have a rotatory motion given them by a steam engine, the shafts of the spirals passing through a water-tight packing into the vessel."

A.D. 1830.—William Church also claims (No. 6041) the application of oscillating cylinders with hollow trunnions to driving paddle wheel shafts and to work expansively by means of a throttle valve operated by means of tappets. Joseph Maudslay, however, in 1827 had also patented an oscillating cylinder engine. Goldsworthy Gurney, in the same year, used oscillating cylinders to drive his steam coach.

PLATE II.

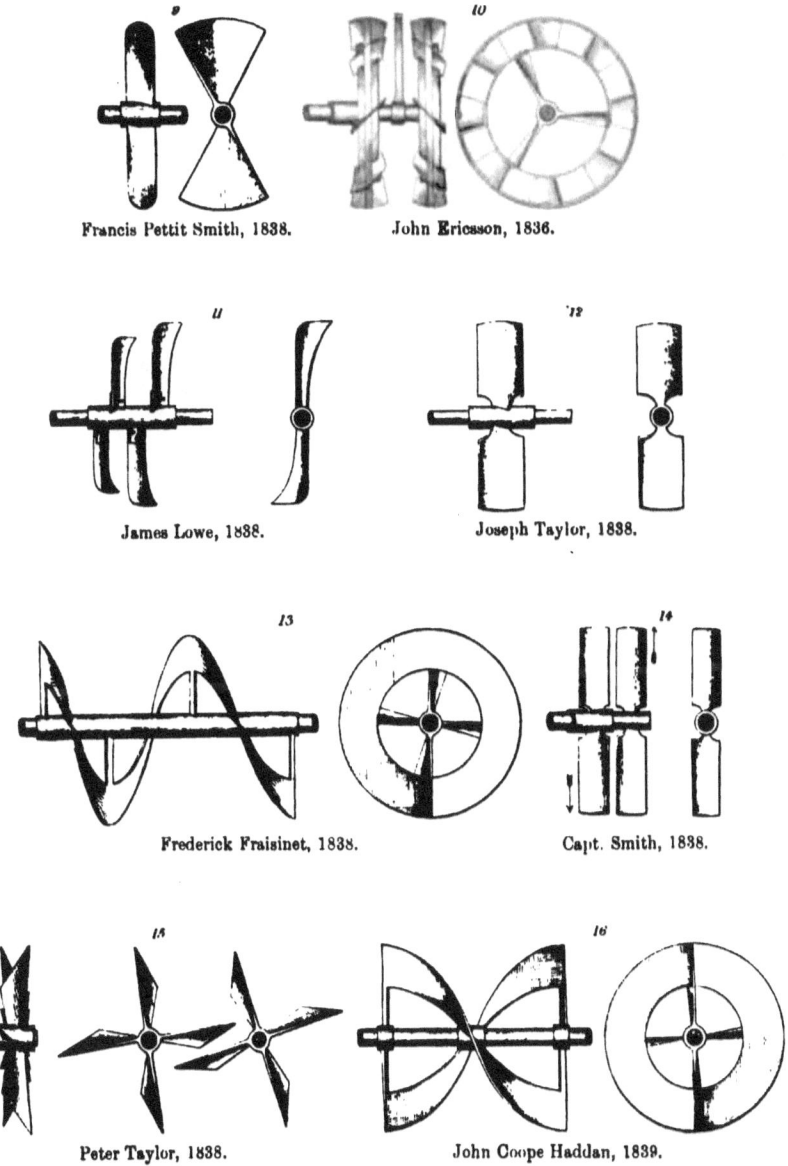

Francis Pettit Smith, 1838. John Ericsson, 1836.

James Lowe, 1838. Joseph Taylor, 1838.

Frederick Fraisinet, 1838. Capt. Smith, 1838.

Peter Taylor, 1838. John Coope Haddan, 1839.

PLATE III.

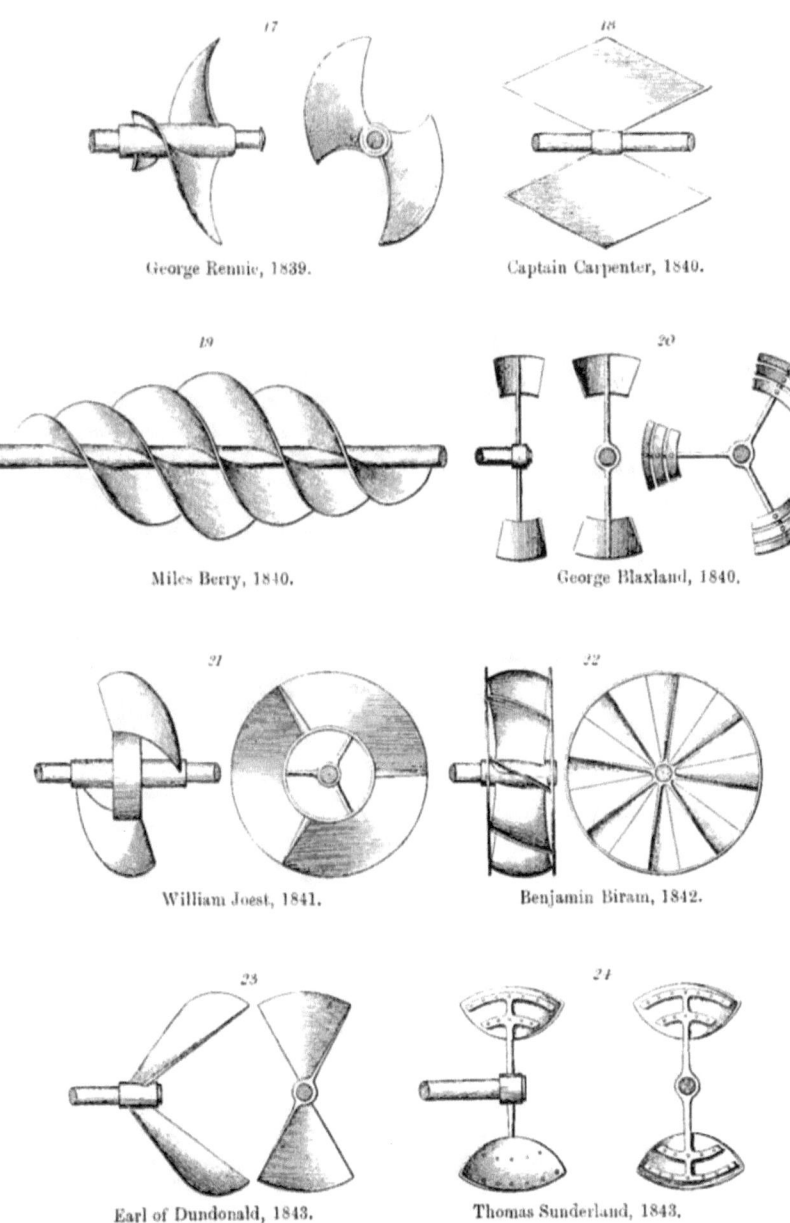

To turn the vessel a valve causes the water to be expelled at one side in direction of the bows. The Admiralty built H.M.S. "Waterwitch" in 1866 to try Ruthven's invention. The results as to speed were very disappointing.

A.D. 1840.—Captain Edward J. Carpenter proposed a patent No. 8545 (see fig. 18, Plate III.) to fit twin screws with shafts having Hook's universal joint to permit of the screw being lifted out of water when under sail, and the dividing of the rudder into two parts connected by a hasp forging when a single screw is used projecting beyond the rudder.

George Blaxland in this year patented the idea of making a screw blade in a series of strips set at different angles so as to constitute steps. (See fig. 20, Plate III.)

In this year the "Archimedes," after competing with the Cross-Channel Dover steamer, made a voyage round Great Britain, calling at all principal ports. (*Vide* Chapter XV.)

A.D. 1841.—David Napier, in patent No. 8893, proposed two paddle wheels with oblique floats or blades like a screw, having their shafts parallel to one another in a fore and aft direction, but above water at the stern, so that the arms were about half immersed. They turned in opposite directions, and had their centres so close that one half masked the other. They were, in fact, overlapping twin screws with their shaft centres a quarter their diameter above water, and so behaved much as a twin screw bluff-sterned ship would to-day under similar circumstances. The screws were in this case, however, in a very bad position for getting a good water supply, and consequently the experiment failed badly. (See fig. 6.)

A.D. 1843.—John Laird took out patent No. 9830.

The screw shaft is surrounded by a watertight trunk, "by which arrangement, cargo may be stowed around and about the trunk, and the steam engine and machinery may be amidships." The screw shaft works through the trunk watertight at both ends.

A.D. 1843.—James Hamer proposed, in patent No. 9592, to drive a screw propeller by means of a turbine on Branca's principle and fit it to a shaft connected to the driving shaft by a Hook's joint.

Joseph Maudslay in the same year (No. 9833) suggested fitting a rudder on each side of a single screw projecting downwards from the quarter without external bearings and having their tillers connected, very much as done in later days by Sir John Thornycroft.

H.M.S. "Rattler," the first screw steamer in H. M. Navy, was completed this year; she was 176·5 feet long, 32·7 feet beam, and 13·5 feet draught of water, 1140 tons displacement, propelled by a screw 10 feet in diameter and 11 feet pitch, driven by Maudslay's twin-cylinder engines of 200 N.H.P. with four cylinders 40 inches diameter and 4 feet stroke. Her speed was 10 knots. (See fig. 7, also Chapter XV.)

A.D. 1844.—Bennet Woodcroft claims to fit a propeller with blades capable of turning round on the axis of their shanks and moved by means of a sliding sleeve within the boss, having on its surface a spiral groove into which the pinion, the arms, or levers of the shanks are fitted, and free to slide and have an angular movement as the sleeve slides axially on the shaft.

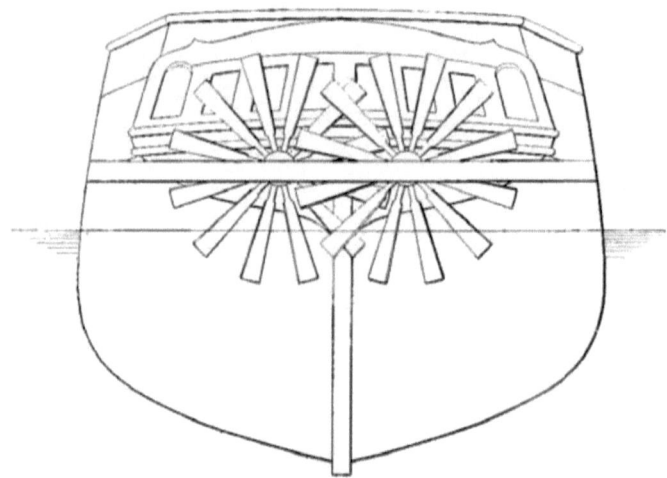

FIG. 6.—Napier's Double Screws, 1891.

A.D. 1845.—The screw steamer "Great Britain," 3270 tons, crossed the Atlantic. She was 322 feet long, 48 feet beam, and 31·5 feet deep, and had engines of 1500 I.H.P. geared to the screw shaft. The four cylinders were 80 inches diameter and 72 inches stroke. This was the first screw steamer and first iron ship placed on this station. She was designed as a paddle-wheel ship, but altered when partly built to a screw by Brunel. She did not, however, remain long on the New York and Liverpool station.

A.D. 1846.—Joseph Maudslay proposed to fit screws that they might be lifted on deck through an aperture. The shaft end is conical, and fits into a conical recess in the boss and tightened up by end pressure.

PLATE IV.

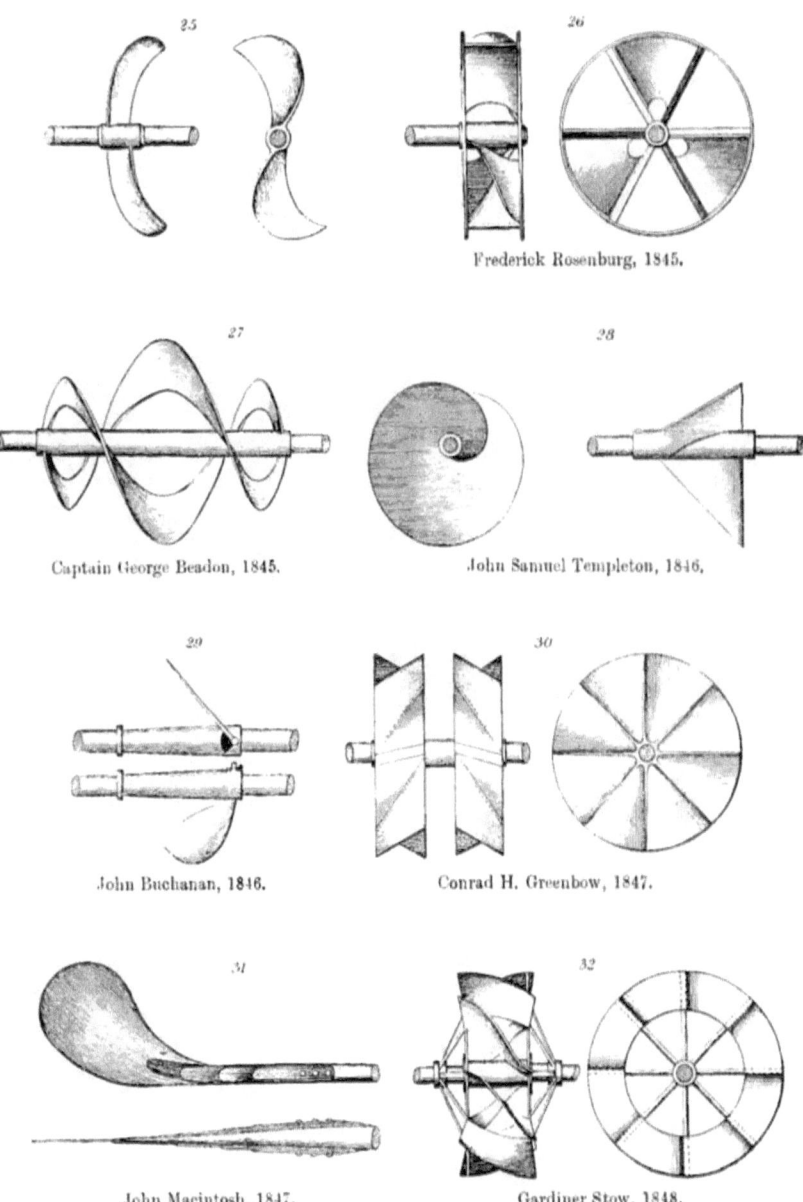

A.D. 1846.—Henry Bessemer claimed (No. 11,352) to use the exhaust steam from a reciprocating engine to turn a hollow shaft, through which it is caused to pass and discharge through two axes having tangential openings; these axes may revolve inside the condenser. This undoubtedly is the rudiment of the low-pressure

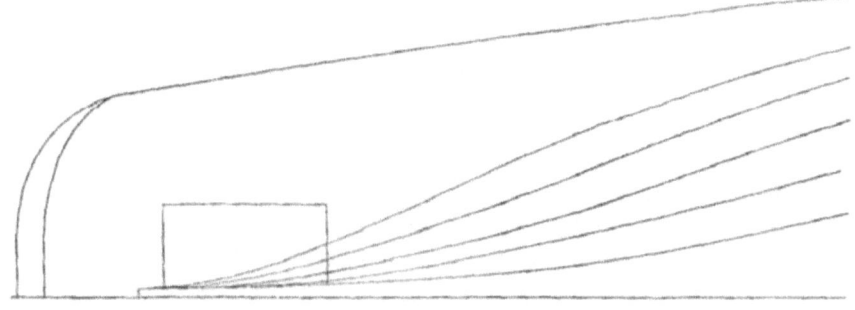

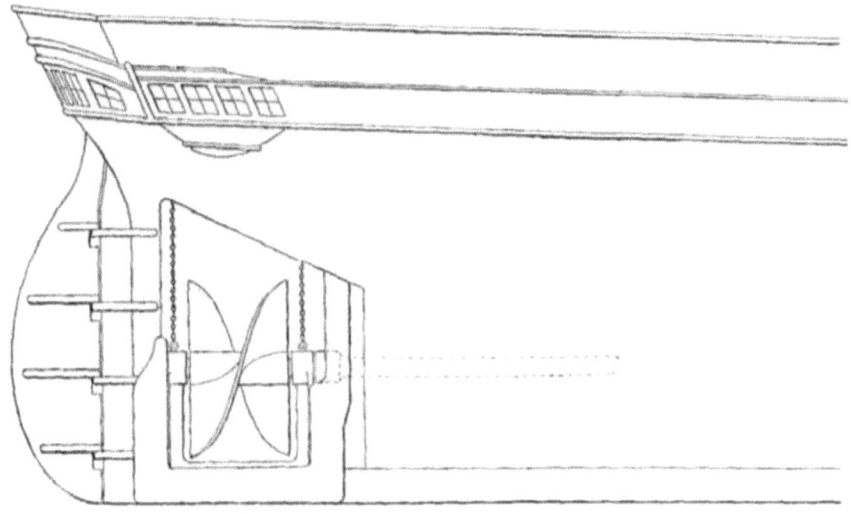

Fig. 7.—Stern of H.M.S. "Rattler," 1843.

turbine; and it is singular that Mr Bessemer with his scientific knowledge did not appreciate the true value of his invention, or rather the adaptation of Hero's reaction engine—the first form of steam engine—when worked in conjunction with a condenser.

A.D. 1847.—John Macintosh (No. 11,763) proposed to make the blades of a screw (see fig. 31, Plate IV.) so that they could be turned sufficiently to act either "ahead" or "astern." Such screws are now often used in small craft propelled by oil and gas engines.

Johann Gottlob Seyrig in this year claimed (No. 11,695) "the use of a turbine or centrifugal wheel and a screw propeller in such conjunction that the currents produced shall simultaneously help both machines," much as Sir J. F. Thornycroft did with the "Lightning" and other ships in 1882.

A.D. 1848.—Joseph Maudslay claimed (No. 12,088) to make screws with the shanks of opposite blades overlapping in the boss and geared together by sectors, so that as the shaft rotates the blades turn into position, and when it stops the water places them fore and aft.

A.D. 1849.—Wakefield Pim proposed (No. 12,440) to fit a screw at the bow as well as stern to work in conjunction.

John Dugdale and Edward Birch claimed (No. 12,625) to fit one or more propellers on one shaft, each being larger as it is nearer the stern.

John Ruthven took out in this year his celebrated patent (No. 12,739) in which he claims to use a centrifugal wheel with curved blades supplied with water admitted through apertures below the ship, and flowing through passages and forced through pipes terminating in nozzles outside the vessel (see fig. 25), the nozzles to be jointed so as to be turned in any direction and the motion of the vessel regulated and directed without stopping the engines.

Robert Griffiths, also in this year, claimed to fit his propellers so that each blade may turn on its axis in a socket by the pressure of the water on the leading side of it, and thus act against a spring (see fig. 8). He expected in this way that the pitch would be increased as the velocity increased, and *vice versâ*, a most valuable feature in those days of auxiliary steamers, as it enabled the engines to be used advantageously when the ship was under sail. Otherwise by a passing squall or other sudden quickening of the ship the engines were liable to "race" to a dangerous extent.

He also specified some distinct forms of blade, one of which is shown fig. 37, Plate V., and claimed to make the blades with their outer ends curved towards the bow, as he made them all eventually. But he also proposed to bend two sternwards and two forwards, as if uncertain as to which was the best direction.

George Callaway and R. A. Purkiss took out a patent this year

PLATE V.

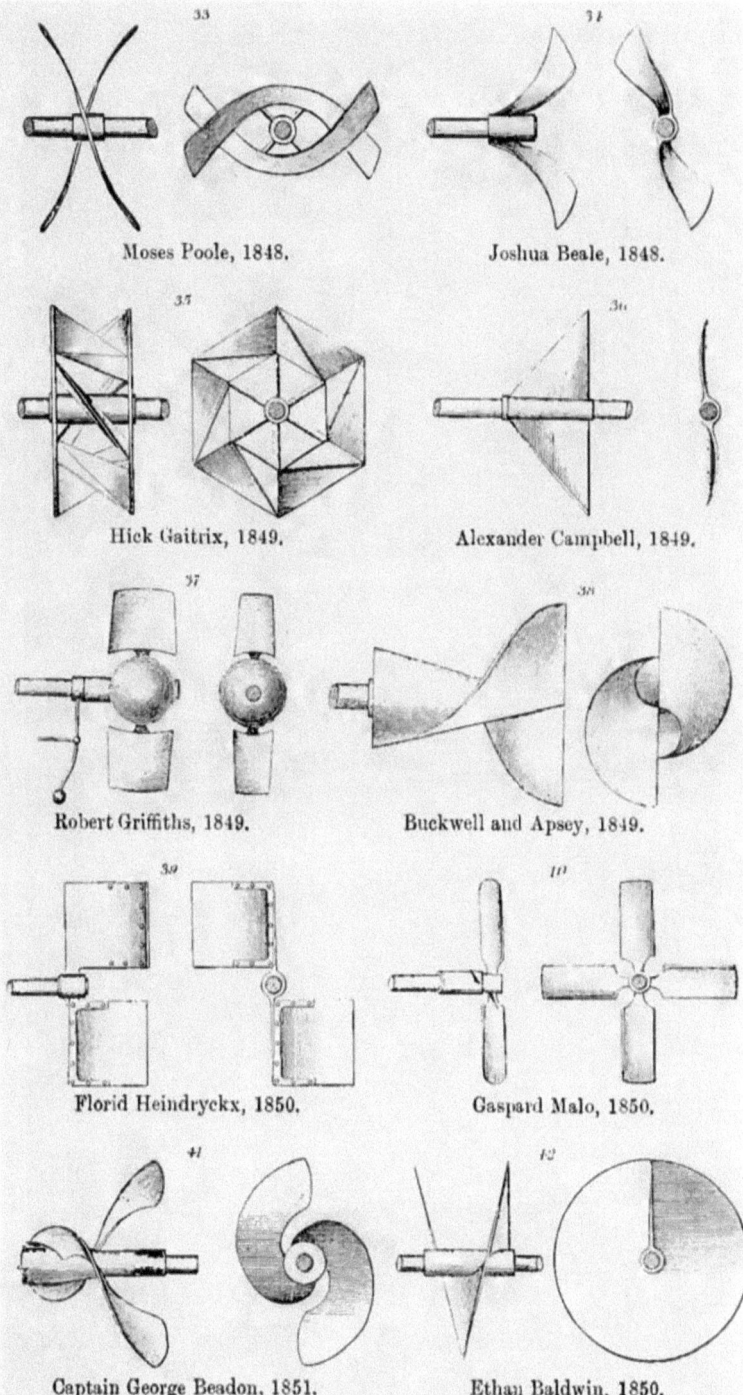

on lines almost identical with those of Ruthven, but three months after him.

A.D. 1850.—Henry Wimshurst proposed, among other things, in his patent No. 13,340, an apparatus for measuring the power exerted in turning the shaft of a screw propeller which would be applicable now to propellers driven by turbines, and was the forerunner of the

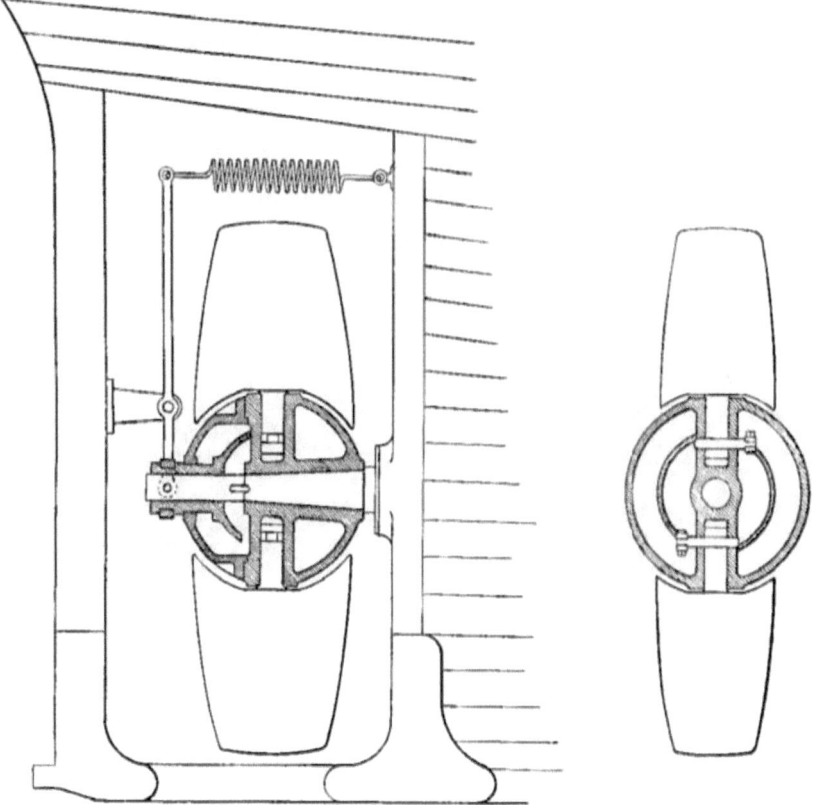

FIG. 8.—Griffiths' Early Patent Screw (Self-adjusting).

Torque instruments at present used to determine the power transmitted by shafts.

A.D. 1850.—The screw steamer "City of Glasgow," 1600 tons, 237 feet long, 34 feet beam, with engines of 350 N.H.P., built of iron by Tod & M'Gregor, Glasgow, was the first of the famous Inman line of steamers. It was placed on the New York station to run against the paddle steamers of the Cunard Company in this year.

A.D. 1851.—Bennet Woodcroft claimed (No. 13,476) to set the

blades of a propeller at any angle by means of toothed wheels and worms, as shown on fig. 9, by a sliding sleeve with spiral groove, as described in his 1844 patent; and further, to move the blades, even to the extent of reversing them, by controlling the apparatus by means of rods fitted in grooves in the shaft and thus carried inboard and worked in the tunnel or engine-room.

Gustav A. Buckholz proposed (No. 13,515) in this year to fit ships with three screw propellers, the middle one being nearer to the stern than the others, all geared together, however, so as to be worked by

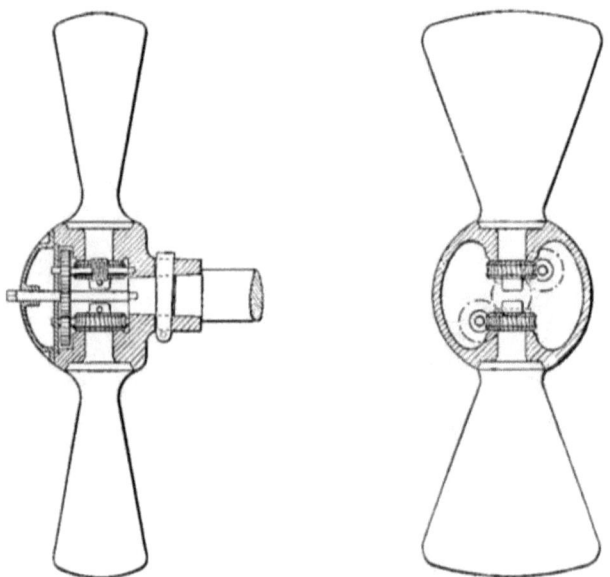

FIG. 9.—Woodcroft's Adjustable Blades.

one engine. In 1855 H.M.S. "Meteor" was fitted in this way, but without success.

A.D. 1851.—Edward J. Carpenter proposed to form twin screw ships with two submerged after bodies, each with aperture, rudder, etc. H.M.S. "Penelope" was built on this plan in 1867, as were also H.M. gunboats "Viper" and "Vixen" by Sir E. J. Reed.

Richard Roberts in this year claimed (No. 13,779) improvements in screw propellers, "making the boss much larger than usual in order that the vanes may act more effectively on the water, and in extending the bosses backwards far enough to admit of their being tapered or otherwise formed so as to allow the water to close upon them without a counter-current being produced." The boss was to be at least one-third the diameter and to have at its after end a curved conical point, and forward the boss was to be "softened off with the body of the vessel." A most valuable improvement, and one much appreciated thirty years after, and since, where high efficiency is requisite with high speeds. (See fig. 10.)

A.D. 1852.—William Clark propounded the making of a screw with the pitch near the tip greater than that at the boss, the latter being "equal only to the speed of the vessel." He did not complete his patent.

Donald Beatson and Thos. Hall in this year patented a screw with a flange at each blade tip so as to diminish the slip by retaining the water.

A.D. 1853.—Beatson claimed (No. 175) to make propeller blades corrugated, ribbed, fluted, or ridged in lines.

Robert Griffiths took out another patent in this year (No. 492),

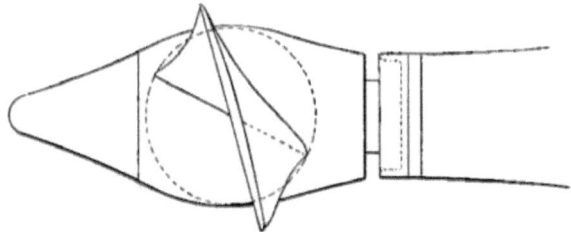

FIG. 10.—Roberts' Patent Boss, 1851.

and includes the spherical boss with gear inside it setting the blades (see fig. 8), but consisting of a pinion fixed to each blade shank geared to a bevelled segment, etc.

Joseph Maudslay claimed, in patent No. 646 of this year, to turn round propeller blades by a lever on the side of the shank worked by a link taken into a grooved sliding collar, etc. (See fig. 47.)

John Fisher proposed making blades with various openings through them; to form flanges or ridges at the back, and to enamel the surface.

James M'Connell claimed, in patent No. 1775, to make screw propeller shafts hollow or tubular, as is now always done for H.M. Navy.

James Mackay proposed to shape the "deadwood" of the ship near the screw into a cylindrical enlargement so as to cause the water to run along without closing in and be again disturbed by the boss. This idea was improved upon many years after by Sir John Thornycroft.

A.D. **1854.**—Rennie of London built a twin-screw steamer for the Khedive of Egypt, 60 feet long, 6 feet beam, and 21 inches draught of water. The screws were 24 inches diameter and made 310 revolutions per minute when the boat was travelling ten knots per hour, driven by disc engines 13 inches in diameter supplied with steam at 45 lbs.; with 60 lbs. it was said that a speed of twelve knots was attained. The following year a similar, but somewhat larger, boat, 70 feet long and 7·5 feet beam, did ten knots at 260 revolutions and consumed 100 lbs. of coal per hour.

In this same year was built the *screw* steamer "Brandon," 210 feet long and 20 feet beam, and was the first ship fitted with compound engines. They were of Randolph & Elder's patent design, having two H.P. cylinders 41 inches diameter and two L.P. 64 inches diameter, working two crank shafts and geared to the screw shaft. The boiler pressure was only 22 lbs. and the cut-off in H.P. cylinder four-tenths. With a cargo of 650 tons on board, she steamed the voyage Glasgow to Limerick at eleven knots on a consumption of 14 cwt. of Scotch coal per hour.

John Penn in this year took out a patent, No. 2114, for fitting fillets of lignum vitæ or some other hard wood in dovetail grooves in the underwater bearings of propeller shafting, now almost the universal practice.

William Wain proposed to operate on the blades of a screw so as to change their angle by a rod down the centre of the shaft, "either by sliding or turning with suitable gear."

A.D. **1855.**—George Peacock claims to make his screw of wrought iron with blades shaped like a bee's wing or parabolic in their curvature.

Henry Bessemer, on the other hand, claimed (No. 1382) to make screw propellers, their shafts and cranks, of "cast steel and pig iron."

Casimir Deschamps and Charles Vilcoq patented (No. 1646) a "free diving boat," what would be called now a "submersible boat," propelled by a screw, etc., and having an *electric light* at the top.

William E. Kenworthy and Henry Greenwood protected the invention of fixing blades to a boss by dovetailing their ends into grooves. This method was years afterwards adopted both by Yarrow and Thornycroft for fixing the blades to torpedo boats and destroyers' propellers. The patent was never completed. (*Vide* fig. 61.)

Christian Schiele proposed a rotary engine for driving screw propellers that was really a turbine, inasmuch as he proposed (No. 1693) to turn the shaft by the action of steam directed against incisions in the circumference of a wheel upon the shaft. The wheel is in a case connected with the condenser. The steam is said to impinge on one side of each of the curved vanes round the periphery, and escape by the open sides into the case.

A.D. 1857.—John Bourne (No. 935) claimed to propel ships by injecting and burning fuel in a fine state of dust in a closed chamber containing air. He said: "The hot air may be used for propelling vessels by being used to move machines resembling a Barker's mill, a smoke jack, or a turbine."

A.D. 1858.—Robert Griffiths (No. 319) proposed to make his screws "so that if a straight edge is held against the blade and perpendicularly to the length of the axis it will be found that at a point at a distance from the axis (generally about half the radius of the screw) the straight edge and the blade part company, the blade falling forward towards the ship, or the screw may be so constructed that the blade shall continually fall away from the straight edge."

A.D. 1858.—The steamer "Great Eastern" was built by John Scott Russell from the general designs of Brunel.

She was 680 feet long, 82·8 feet beam, and 48·2 feet deep. Her gross register tonnage was 18,915 tons, and the displacement 32,000 tons. She was propelled by a single screw and a pair of side wheels. The screw engines had four cylinders 84 inches diameter and 48 inches stroke. The paddle engines had four cylinders 74 inches diameter and 174 inches stroke.

The only other instance of a ship having paddles as well as a screw was the "Bee," a small ship used for instructional purposes at Portsmouth College in the early fifties of the ninteenth century.

A.D. 1859.—Thomas Symons proposed to fit two propellers, "one *above* the other, each of them being of less diameter than the solitary one in present use, and may be driven at a greater speed without producing the like vibration; and the risk of their fouling by floating materials or being struck by shot will be greatly diminished."

A.D. 1860.—Herman Hirsch took out patent No. 2930, in which he states: "The improvements apply to the form of the blades of propellers for vessels, the surfaces of which are made of such a curvature that if sections were made by cylindrical surfaces concentric with the axis of the propeller, the profiles of the sections on these cylindrical surfaces would show a pitch gradually increasing from the entering edge in such a manner that every successive portion of the surface, reckoned from the entering edge, gives the water an additional impulse backwards as it revolves through it; and if sections were made by planes perpendicular to the axis of the propeller the profiles of the surfaces on these planes would be spiral lines concave towards the water, acted on in such a manner that every portion of the surface in revolving through the water impels it to some extent towards the axis, and thus overcomes a portion of its centrifugal tendency. That surface of the blade which acts on the water is therefore made of such a curvature as to combine the two curvatures above described, namely, the curvature of increasing pitch as projected on cylindrical sections, and the spiral curvature as projected on planes perpendicular to the axis." (Fig. 11.)

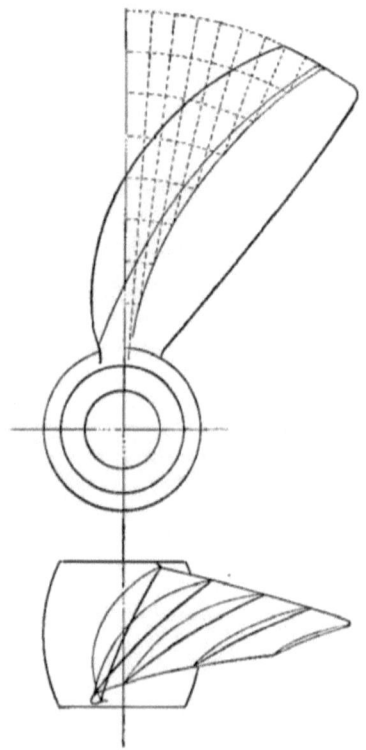

Fig. 11.—H. Hirsch's Screw Propeller of 30 Degrees.

A.D. 1860.—Robert Griffiths took out a fresh patent, No. 2976, in which he states: "This invention has for its object improvements in screw propeller blades which decrease in their width of surface as they become more and more distant from the propeller shaft. It is preferred that each screw propeller blade should be a portion of a true screw of the pitch desired, excepting at the further edges of the blades, which after edges are each composed of an angular surface which is in its whole length at the same or nearly the same angle to the propeller shaft as that at which the widest part of the screw propeller blade stands to the shaft." The widest part of the blade should be "at a point about one-half the radius of the screw from the centre of the propeller shaft," but its position may be varied. (See fig. 38.)

"The angular surface at the edge of the blade commences at or springs from that part of the screw propeller blade which is widest, and it becomes wider and wider as it proceeds outwards to the periphery or circumference of the propeller blade; the angular surface stands at an inclination to the after face of the propeller blade, and consequently, as the propeller blade rotates, the water which has been acted on and put in motion by the fore part of the blade is again struck by this after portion of the blade."

"Screw propellers of smaller diameter than those heretofore used may be employed without decreasing the propelling effect."

The last paragraph is very interesting, especially in the light of modern experience.

A.D. 1861.—Robert Wilson patented, among other things, the idea of making the section of a propeller "a long and narrow ellipse with pointed extremities so as to cause little disturbance in passing through the water."

A.D. 1861.—William Holland Furlonge claimed "in screw steamers, the propeller can, if desired, be made to draw the water through the condenser and deliver it near the stern post."

A.D. 1862.—Alfred Krupp, in his patent No. 1116, says the "invention consists in forming screw propellers in one piece or in two or more pieces from a solid block or blocks of cast steel, and forging the said block or blocks into the necessary shape."

A.D. 1862.—Robert Griffiths extended his patent by taking out No. 1618, and containing the following:—"The improvements consist, first, in constructing screw propellers for steam ships and boats with blades and centre boss of similar form and construction (or the blades may be cast on the boss) to those described in the Specification of a Patent granted to me the 20th February 1858, No. 319, but having four blades (or two sets of blades) which are to be fixed either to the same boss or to separate bosses on to the screw shaft, and so fixed that one set (or pair of blades) is placed before the other set, the first set or pair of blades next the shaft to be of larger diameter than the after set, so as to get greater hold on the water for propelling the ship."

A.D. 1862.—Thomas Carvin patented No. 2301, and claimed to make the propeller blades in the shape "of an elongated irregular oval."

A.D. 1862.—The last large paddle steamer for the Atlantic trade, the "Scotia," was launched and owned by the Cunard Company. This ship in 1879 was converted to a twin screw, and employed after as a cable-laying and repairing ship. She was 379 feet long, 47·8 feet beam, and 37·7 feet deep.

A.D. 1863.—Arthur Rigg, junior, claimed "to surround a screw propeller by a cylinder and to have a grating in front of it to prevent weeds entering."

A.D. 1863.—F. E. Sickels proposed to make "propeller blades of vulcanite strengthened with iron bands."

A.D. 1864.—William B. Adams took out protection for the use of "liquid fuel, such as oil, melted grease of any kind," but he preferred coal oil, petroleum, or shale oil for the propulsion of vessels, and described the burners by which it might be used.

A.D. 1866.—Herman Hirsch obtained another patent, No. 17, in which he claimed: "In the improved propeller the front or entering edge is so inclined to the axis that it cuts the unbroken water with little or no resistance, and the rest of the blade gently curves backwards, being more and more inclined so as to give the water a gradually increased backward motion, thus not only avoiding the excessive resistance caused by a sudden impulse on the water, but also maintaining an uniform reactive pressure from the unbroken water over the whole breadth of the revolving blade."

"The two extreme lines and the whole of the intermediary lines of the surface are spirals comprised in an angle of 60 degrees, thus generating from the axis to the circumference a hollow curved or spoon-shaped form of blades." "The points of the spirals may be rounded off," and their curvatures modified near the axis, to give additional strength. (See fig. 40.)

A.D. 1866.—John Henry Johnson patented No. 256, and proposed to use two distinct sets of paddle wheels, the after pair about one-eighth larger than the forward pair; and he goes on to say that in a ship 500 feet long the two sets of wheels are to be about 150 feet apart.

A.D. 1866.—William Dudgeon, in his patent No. 2068, claimed among other things that "as regards ships, the object sought is to provide support for the two screw shafts, which project astern in parallel lines with the keel from under the ships' quarters, and respectively are encased and carry the screws. For this purpose a flat horizontal chamber is constructed on each side and fixed to the skin of the vessel, projecting laterally therefrom so as to enclose and support the screw shafts; and inside the vessel, extending from side to side, is fixed a horizontal plate framing to correspond with the level of the shafts, so as to form a substantial lateral support between the chambers, which are rendered more firm by a transverse vertical bulkhead. The extreme ends of the chambers, wherein the shaft bearings and stuffing-boxes are fitted, are further supported by wrought-iron diagonal stays or brackets, which are fixed to the sides of the vessel by riveting through the skin and ribs."

A.D. 1866.—Dr A. C. Kirk took out provisional protection for the following: "This invention has principally for its object to render steam dredgers capable of being more easily manœuvred than hitherto; and it consists in employing for that purpose centrifugal or other pumping apparatus to be worked by the main or separate engines, and to cause the projection by suitable passages or orifices at the stern of one or more streams of water."

"As it is of great importance to have the power of turning and generally manœuvring such vessels independently of the tug usually in attendance, deflectors or rudders are attached to the stern orifices through which the water is projected; whilst to permit of reversing the direction of propulsion, either rotatory reversible pumps are used, or reversing valves are fitted to the water passages communicating with the pumping apparatus."

A.D. 1866.—J. S. Martin and J. F. Droop claimed to propel and steer ships by means of a stream of water ejected by means of a jet of steam, as in the case of the Giffard injector.

A.D. 1867.—Robert Atkin claimed to fit three, four, five, and six screw propellers; namely, two at each end and at any suitable distance from the stem and stern.

A.D. 1873.—James Howden took out patent No. 3278, in which the steamer is fitted with a large propeller at each end, "the propeller being of such a size that the blade thereof extends below the keel." The keel may be curved downwards at the bow and stern so as to pass under the propeller.

A.D. 1873.—Samuel Osborne and Stephen Alley took out patent No. 3379. The invention consists in forming propellers of thin sheets of metal—steel, gun-metal, etc. Each blade may be made of a single sheet, but it is preferred to be made of two thicknesses with a space in the middle. The blades thus formed are fixed on the boss in various ways. They may be let into a groove in the boss and fixed by wedges, or flanges on the blade may be bolted to the boss, or the blades may be fitted to flanged pieces in sockets in the boss, or the boss may be made up of four quadrants the flat surfaces of which are shaped so as to receive the blades between them.

A.D. 1873.—John Isaac Thornycroft took out patent No. 3551. The object of the invention is to cause the water to be driven backward by the screw, "in what may be termed hollow cylinders, or concentric annular volumes." For this purpose each blade is curved in the direction of its length, "in such manner that assuming the blade to be cut in a plane passing through the axis of the propeller, and through the centre of the blade in a direction parallel to the said axis, the section of the blade will be convex on its driving face, that is, on the outer, or, in other words, the after side or surface," the pitch of the blades increases gradually from the boss towards the centre, and then decreases towards the outer edge in the direction of the screw's diameter. In the direction of the axis the pitch increases towards the after end of the screw.

A.D. 1873.—Robert Griffiths took out patent No. 3817, relating to methods of affording a larger supply of water to screws working in tunnels at the bows and stern.

(1) The tunnel for a stern propeller is made with a "lip, scoop, or projection," extending down from the after side of the mouth of the tunnel. The projecting piece may be hinged, so that it can be raised up when not required, as in the case of vessels driven by screws both at the bows and stern, in which case it is stated that the additional speed given by the bow propeller ensures a sufficient supply of water to the propeller at the stern.

(2) The tunnel for the bow propeller is made with a flaring mouth for the above-mentioned purpose.

(3) Similar arrangements may be made when twin screws are employed.

(4) When one screw only at the stern is used, it is preferred to make it "of parallel form and of increasing pitch from end to end." With such a screw the "lip" is not required.

(5) A propeller "of increasing pitch and taper form" is used in the stern tunnel.

(6) A screw of increasing pitch is mounted on a taper boss, the screw itself being "parallel from end to end."

(7) The outlet end of the stern tunnels may be widened to allow the water to escape freely.

A.D. 1874.—Sir J. I. Thornycroft took out patent No. 382. A propeller is placed "within an external recess or recesses so formed in the under part of the hull that when the vessel is afloat the mouth or opening, or mouths or openings, of such recess or recesses shall be below the surface or level of the water in which the vessel is floating, and the crown or crowns of the recess or recesses considerably above the said surface or level." The action of the propeller empties this chamber of air and fills it with water which rises above the level of the external water, or an air pump may be employed for the purpose. (See fig. 12.)

A.D. 1874.—Sir E. J. Reed took out patent No. 2565. The invention consists in using two screw propellers in the same axial line, one being on a sleeve and the other on a shaft within the sleeve. One of the propellers is larger than the other. In deep water both are to be employed; in shallow water the smaller propeller only, the larger

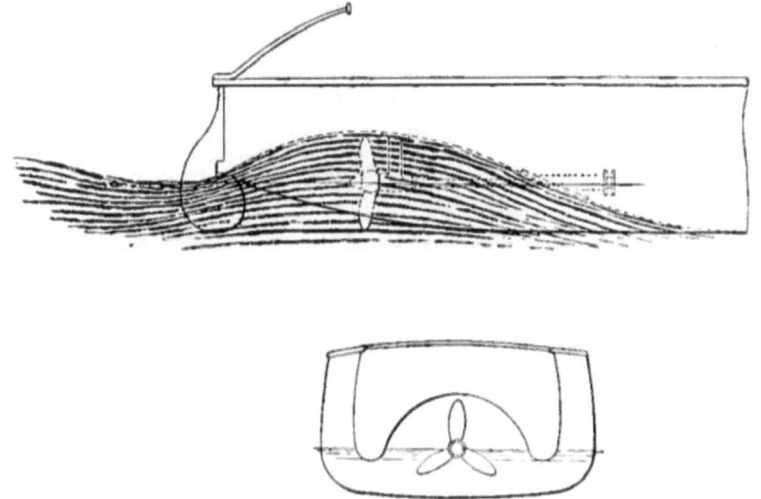

Fig. 12.—Thornycroft's Stern for Shallow Draught Screw Ships.

one being two-bladed and placed horizontally so as to be clear of the bottom.

A.D. 1874.—James Howden took out patent No. 3246 for composite screw blades. The blade is formed of one or more plates of sheet iron riveted to a stem formed with cross pieces, this stem occupying the middle line of the blade. One plate may form the front and the other the back of the blade, or one plate alone may be employed, filling pieces being fitted to the back. "The propelling face of the blade may also have one or more parallel bars of a limited width riveted or otherwise fastened through the plate, stock, and filling pieces, so as to further stiffen the blades, these bars being curved and fixed on the face of the blades to the circle of the revolution of the propeller. The central bar of the stock may also be increased across the back of the plate or sheet forming the face of the blade, so as to take the place of the filling pieces."

A.D. 1875.—Joseph Hirsch took out patent No. 576, being improvements on No. 2930, A.D. 1860, and No. 17, A.D. 1866.

The invention refers to a special shape to be given to the propeller blade. The entering edge of the blade is slightly curved and the opposite edge has " a considerably greater curvature," the lines of the intermediate sections of the blade "being more and more curved as they are nearer to the leaving edge. This increase in forward curvature is determined in such a manner that the blade has a pitch increasing from the front edge backwards; but this increase of pitch is greatest at the tip of the blade and becomes less and less towards the root. It is preferred that the pitch along the entering edge should be that due to the forward motion of the vessel, so that the blade has there an obliquity to the axis determined by compounding the velocity of rotation with the velocity of advance, and the increase of pitch at the tip of the blade may be from 20 to 30 per cent." The blade can be made with "a uniform width from the root to the tip," or it may be made "to taper in width from the root outwards, the intersections on its face being in that case not planes perpendicular to the axis, but conical or conoidal surfaces having an obliquity to the axis which becomes less towards the middle of the blade's width."

A.D. 1875.—Hermann Hirsch took out patent No. 4479 for " Improvements in screw propellers." The invention relates to making the propeller blade of one of two forms. In the first form the blade has " a double curve, so that seen in edge view in a longitudinal direction, it will present the figure of an elongated S." In the second form it has " two such double curves in juxtaposition, so that the profile formed by its opposite edges will, when seen in an oblique direction together, present the figure of 8 more or less elongated."

Yarrow's Shallow Draught Screw Arrangement.—Some very interesting experiments were made by Mr Yarrow with a view to adopting the propeller as a means of driving shallow draught ships whose draught of water would be considerably less than the diameter of the screw. Griffiths had suggested a method of doing this, and Mr, now Sir John, Thornycroft, carried out the idea most successfully on the plan shown in fig. 12. But Mr Yarrow wished to go a step further, and to see how far it would be of advantage in such ships if

MODERN HISTORY OF PROPELLERS.

the obstruction abaft the screw were removed when the ship was in deeper water, and did not require the compulsory submersion of the screw. For this purpose he fitted a flap or shutter hinged to the roof of the waterway in wake of the screw, and arranged to be raised or

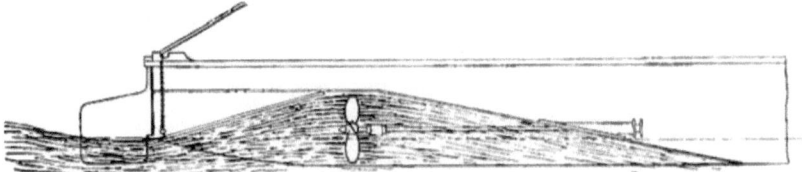

Flap-down Shallow Draught.

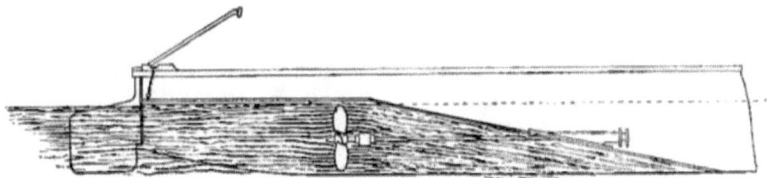

Flap-up Deep Draught.

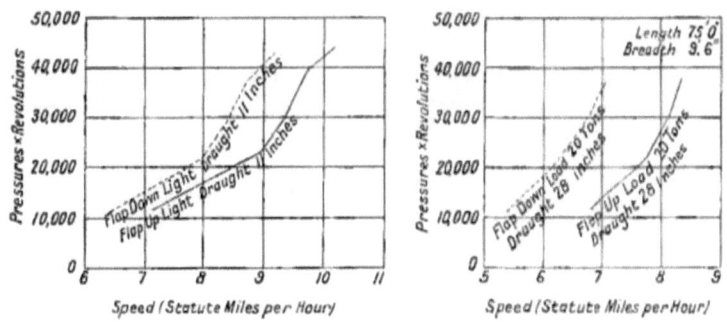

FIG. 12A.—Yarrow's Drop Flap for Shallow Draught Screw Ships.

lowered to suit the draught of water of the boat. That is to say, the flap could always be lowered down so as to touch the water and thereby seal the screw race from the air. The results of the trials made by Yarrow with a boat fitted with this arrangement at the stern is clearly set forth in fig. 12A.

CHAPTER III.

RESISTANCE OF SHIPS.

IF a plane or thin sheet of material is pushed or drawn through the water in a direction at right angles to its surface, it resists very vigorously; if it is moved in the direction of its surface the resistance, though very slight, is appreciable, especially if its surface is rough.

It will be observed in the former experiment that the water in front has to swell up against the plane and pass around it on either side; behind the plane there is a tendency to form a depression or space free of water, especially if the movement is accelerated; and further, that the water in flowing around the sides of the plane and from the bottom fills up the space behind with confused motion and the formation of eddies. If the speed be further quickened it will be observed that there is a flow of water from behind approaching the plane in the direction in which it is moving, due to gravity acting on the water to make it fill the now increasing cavity behind the plate.

So far back as 1798 Colonel Beaufoy made some experiments with such planes deeply immersed with a view to ascertain the exact amount of resistance at various speeds and positions, and found

$$\text{The resistance } R = f \frac{\sigma}{2g} \times Av^2,$$

where A is the area in square feet, v the velocity in feet per second, σ the weight of a cubic foot of liquid, g gravity ($=32$).

Now f is the factor which Beaufoy set out to determine, and which he finally concluded had a value of $1\cdot 1$.

Hence for sea water with $\sigma = 64$, resistance $= 1\cdot 1 \times Av^2$.

Many years after, however, Dr W. Froude was led to put quite a different value, viz. 1·7; and later Dubuat estimated it at 1·43, while another experimenter made it 1·6. It is now very interesting to know that Mr R. E. Froude, after most careful and exhaustive experimenting, has come to the conclusion that Beaufoy was right in putting the value of f at 1·1.

While the value of f has been confirmed, doubts have been cast on the resistance varying exactly with the area, or exactly as the square of the velocity. It will be seen later on that the resistance of ships in practice varies with a lower velocity index than 2, except when the bottom is rough as with fine sand cast on the paint.

It is, however, sufficient for all purposes here to assume generally that R does vary with A and v^2.

If the plane be bent into a curve and towed with the convex face leading, the resistance will be considerably reduced; if the plane be bent so as to form an angle, the resistance will also be comparatively small, and be somewhere between R and the skin resistance R.

Now if the plane move in a direction inclined to its surface at an angle θ, then, in sea water it is found by experiment that

$$\text{Resistance } R_1 = 1\cdot 622 \frac{\sin^2 \theta}{0\cdot 39 + 0\cdot 61 \sin \theta} \times Av^2;$$

and the resistance at right angles to the line of motion

$$R_2 = \frac{0\cdot 39 + 0\cdot 61 \sin \theta}{\sin^2 \theta} R^1.$$

Then $R_1 \div R_2 = \sin^2 \theta \div 0\cdot 39 + 0\cdot 61 \sin \theta$.

All ships and bodies with ship-like form in moving through the water are subject to resistance from the friction of the water on the skin or surface submerged, however smooth it may be, and a head resistance, due to the pressure of the body on the water in front; at the stern or following end there is a further cause of loss or increase of resistance, due to the imperfect action of the water in filling in the void space behind, so that it fails to follow up the body without "loss of head"—that is, without any decrease of hydraulic pressure.

A "balk" or log of timber being towed through the water may be observed with advantage, as all these phenomena can be clearly seen and fully appreciated. The pressure at the bow sets up waves by raising a mass in front of it, which only disperses to allow more to form. The friction of the skin at the bow and sides sets in motion other portions of the surrounding water, and from ripples at the sides waves of another kind are formed and spread out fanwise. At the stern further waves are caused by the replacement of the water displaced, and eddies set up in the wake.

Experience has taught raftsmen to tow balk timber with the big end leading, as thereby the pressure is relieved from the sides and the speed at which it moves cannot cause much end resistance or wave-making; also the stern replacement, etc., is easier and effected with less loss of energy.

A ship has not a flat end like a balk; hers are more or less of wedge form; but however fine the entrance may be, a speed is arrived at when definite waves are formed near the bow and flow away from it; that is, the water displaced has not time to spread gently and so be unobservable, but is heaped up more or less: they are called the *waves of displacement*, just as those near the stern caused by the inflow in wake of the ship are called *waves of replacement*, and both are *waves of translation*, caused by the ship being translated from one spot to another.

Skin resistance, or, as some prefer to designate it, "tangential resistance,"

$$R_0 = j \times A v^n,$$

A being, as before, the area in square feet; v the velocity in feet per second; j a factor deduced from experiments by W. Froude, at the same time that he found the value of the index n as follows:—

TABLE I.—FROUDE'S VALUES OF j, WITH INDEX FOR VARIATION 1·825.

Length.	Coefficient Resistance.	Length.	Coefficient Resistance.	Length.	Coefficient Resistance.	Length.	Coefficient Resistance.
Feet.		Feet.		Feet.		Feet.	
80	·00933	140	·00911	250	·00897	450	·00883
90	·00928	160	·00907	300	·00892	500	·00880
100	·00923	180	·00904	350	·00889	550	·00877
120	·00916	200	·00902	400	·00886	600	·00874

TABLE II.—FROUDE'S VALUES OF n AND (j) FROM EXPERIMENTS WITH SURFACES OF DIFFERENT MATERIAL.

Nature of Surface.	Length of Surface, or Distance from Cutwater, in Feet.											
	2 Feet.			8 Feet.			20 Feet.			50 Feet.		
	A	B	C	A	B	C	A	B	C	A	B	C
Varnish . .	2·00	·41	·390	1·85	·325	·264	1·85	·278	·240	1·83	·250	·226
Paraffin . .	1·95	·38	·370	1·94	·314	·260	1·93	·271	·237	...	...	...
Tinfoil . .	2·16	·30	·295	1·99	·278	·263	1·90	·262	·244	1·83	·246	·232
Calico . .	1·93	·87	·725	1·92	·626	·504	1·89	·531	·447	1·87	·474	·423
Fine sand	2·00	·81	·690	2·00	·583	·450	2·00	·480	·384	2·06	·405	·337
Medium ,, .	2·00	·90	·730	2·00	·625	·488	2·00	534	·465	2·00	·488	·456
Coarse ,, .	2·00	1·10	·880	2·00	·714	·520	2·00	·588	·490	...	...	...

Column A is the index or value of n.

Column B gives the *mean* resistance in pounds per square foot of the whole surface to the point named at a speed of 10 feet per second.

Column C gives the *actual* resistance per square foot *at* the distance named for the same speed.

The resistance decreases as the distance from the bow increases, for in the case of varnish, while 2 feet from the bow, it is ·390; 8 feet from the bow, ·264; 20 feet from the bow, ·240; and 50 feet from it is only ·226 lb. per square foot at 10 feet per second.

It was usual formerly to assume that the resistance of a ship varied throughout as the square of the speed, and to state that the I.H.P. varied with the cube of the speed. One result of the early progressive trials was the dissipation of that idea. No doubt, theoretically, it was not far wrong to say that power varied as the cube, but what power was meant is another matter; one thing is certain, it could not be the gross I.H.P. With modern marine engines at full speed the efficiency is from 0·90 to 0·95, so that the amount absorbed in overcoming friction working the pumps, etc., is from 5 to 10 per cent. Of this amount $\frac{2}{3}$ths vary with the revolutions and $\frac{3}{3}$ths with the square of the revolutions. Hence an engine whose efficiency is 90 at full speed at half the revolutions will absorb only 3·5 per cent. of the original I.H.P., while the I.H.P. at half revolutions will be probably one-seventh the gross; the efficiency then will be 0·755.

With turbo motors the mechanical efficiency is still higher, and may be taken at 98 per cent. at full speed.

The wetted skin is, to those who have no model and tank apparatus wherewith to experiment, still the criterion of quantity with which they have to deal and guide them in estimating the power required to drive a ship or the resistance encountered in towing a ship at a certain speed. For, although the wave-making absorbs a large amount of the developed power, and especially so at high speeds, it is trivial with ships when moving at speeds well under the designed full speed. With a tug-boat running free at full power, the waste in wave-making is enormous; with a 20-knot ship moving at 16 knots the waves are inconsiderable and at 10 knots are negligible. At a speed of 66 per cent. of the designed full speed, it is sufficient for practical purposes to assume the skin resistance with a liberal allowance of coefficient of friction as the gross resistance of the ship when towed.

The area of the surface of the ship immersed can of course be

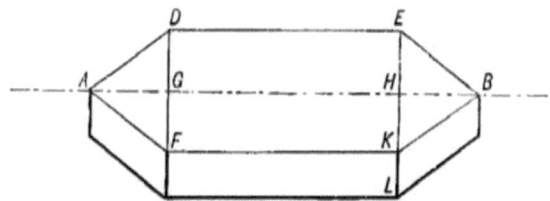

Fig. 12B.—Kirk's Block Model.

measured and computed; but it is a long and troublesome task compared with other methods which give a close approximation to the true amount. Of these the earliest and best known was:—

1. The general idea proposed by Dr Kirk is to reduce all ships to so definite and simple a form that they may be easily compared, and that the magnitude of certain features of this form shall determine the suitability of the ship for speed, etc. As rectangles and triangles are the simplest forms of figure, and more easily compared than surfaces enclosed by curves, so the form chosen is bounded by triangles and rectangles.

The form consists of a middle body, which is a rectangular parallelepiped, and the fore body and after body prisms having isosceles triangles for bases; in other words, it is a vessel having a rectangular midship section, parallel middle body, and wedge-shaped ends, as shown in fig. 12B.

This is called a *block model*, and is such that its length is equal to that of the ship, the depth is equal to the mean draught of water, the capacity equal to the displacement, and its area of section equal to the area of immersed midship section of the ship. The dimensions of the block model may be obtained by the following methods :—

Since AG is supposed equal to HB, and DF equals EK, the triangle ADF equals the triangle EBK, and they together will equal the rectangle whose base is DF and height AG. Therefore, the area ADEBKF equals EK × AH. The volume of the figure is this area multiplied by the height KL. Then the volume of the block is equal to KL × EK × AH. But KL × EK is equal to the area of mid section, which is by supposition equal to the area of immersed midship section of the ship, and the volume of the block is equal to the volume displaced by the ship. Hence

Displacement × 35 = immersed midship section × AH ;

or,

AH = displacement × 35 ÷ immersed midship section.

Now

HB = AB − AH, and AB = the length of the ship.

Therefore, the length of fore-body of block model is equal to the length of the ship, less the value of AH as found above.

Again, the area of section KL × EK is equal to the area of immersed midship section, and KL is equal to the mean draught of water. Therefore

EK = immersed midship section ÷ mean draught of water.

Dr Kirk also found that the wetted surface of this block model is very nearly equal to that of the ship; and as this area is easily calculated from the model, it is a very convenient and simple way of obtaining the wetted skin. In actual practice, the wetted skin of the model is from 2 to 5 per cent. in excess of the ship; for all purposes of comparison and general calculation, it is sufficient to take the surface of the model.

The area of bottom of this model = EK × AH.

The area of sides = 2 × FK × KL = 2 (AB − 2 HB) × KL = 2 (length of ship − 2, length of fore-body) × mean draught of water.

The area of sides of ends = 4 × KB × KL = 4 $\sqrt{HB^2 + HK^2}$ × KL = 4 $\sqrt{\text{Length fore-body}^2 + \text{half breadth of model}^2}$ × mean draught of water.

The angle of entrance is EBL; EBH is half that angle; and the tangent EBH = EH ÷ HB.

Or, tangent of half the angle of entrance = half the breadth of model − length of fore-body.

From this, by means of a table of natural tangents, the angle of entrance may be obtained.

The block model for ocean-going merchant steamers whose speed is from 15 knots upwards has an angle of entrance from 15 to 24 degrees, and a length of fore-body from 0·3 to 0·36 of the length.

That of ocean-going steamers whose speed is from 12 to 15 knots has an angle of entrance from 24 to 30 degrees, and fore-body from 0·26 to 0·3.

Seaton's Rule for angle of entrance of "block model":

$$\text{Angle in degrees} = 70\frac{\sqrt[4]{L}}{S}.$$

L is the length of ship in feet; S is the speed in knots.

Dr Kirk measured the length from the fore side of stem to the aft side of *body-post* on the waterline. This is an unnecessary re-refinement when screw steamers alone are being compared, as then the length may be taken as that "between perpendiculars." However, when small or moderate size screw steamers are being compared with paddle-wheel steamers, it may be necessary to measure in this way.

2. **Mumford's method** is a simple one, and gives results much nearer the actual surfaces than Kirk's.

L is the length of the ship between perpendiculars in feet.

B is the greatest beam; H the depth of immersed midship section.

b is the block coefficient. D is the displacement in tons of 35 cubic feet. Then

$$\text{Wetted skin} = L\,(1\cdot7\,H + bB).$$

3. **Seaton's Rule.**—It is, however, not always easy to ascertain the moulded draught of every ship, for some have flat keels and some bar, the latter of uncertain depth; moreover, while the displacement of a ship on trial is given, and from Lloyd's Register and other sources the length and beam can be always obtained when it is not stated, the draught of water is seldom given. For these reasons, and the fact that quite as accurate results are given by it in even less

RESISTANCE OF SHIPS.

time than by Mumford's, the author has devised and recommends for use the following formula:—

D being the displacement in tons and $K = L \div B$. $F = 42 \times \sqrt[4]{K}$.

$$\text{Wetted skin} = F \times D^{\frac{2}{3}}.$$

For ships 8 to 10 beams $F = 71$ to 74.
    „   6 to 8 beams $F = 67$ to 71.
    „   4·5 to 6 beams $F = 62$ to 67.

The following table gives the actual values of F for the variations in K.

K being the ratio of a ship's length to her beam.

F the factor with which to multiply $D^{\frac{2}{3}}$ being $42 \times \sqrt[4]{K}$.

TABLE III.

K.	F.	K.	F.	K.	F.	K.	F.
4·0	59·40	6·0	65·7	8·0	70·6	10·0	74·7
4·1	59·85	6·1	65·98	8·1	70·82	10·1	74·89
4·2	60·25	6·2	66·24	8·2	71·05	10·2	75·08
4·3	60·6	6·3	66·5	8·3	71·27	10·3	75·24
4·4	61·0	6·4	66·75	8·4	71·48	10·4	75·42
4·5	61·3	6·5	67·0	8·5	71·69	10·5	75·58
4·6	61·65	6·6	67·25	8·6	71·90	10·6	75·85
4·7	61·95	6·7	67·5	8·7	72·11	10·7	76·05
4·8	62·3	6·8	67·75	8·8	72·33	10·8	76·0
4·9	62·6	6·9	68·0	8·9	72·55	10·9	76·18
5·0	62·9	7·0	68·25	9·0	72·76	11·0	76·35
5·1	63·2	7·1	68·5	9·1	72·97	11·1	76·53
5·2	63·5	7·2	68·74	9·2	73·18	11·2	76·68
5·3	63·78	7·3	68·98	9·3	73·39	11·3	76·85
5·4	64·1	7·4	69·23	9·4	73·59	11·4	77·03
5·5	64·35	7·5	69·45	9·5	73·77	11·5	77·2
5·6	64·65	7·6	69·7	9·6	73·96	11·6	77·35
5·7	64·9	7·7	69·93	9·7	74·15	11·7	77·5
5·8	65·18	7·8	70·15	9·8	74·34	11·8	77·65
5·9	65·46	7·9	70·4	9·9	74·53	11·9	77·82

Kirk's Method, when applied to ships of the older type having a good "rise of floor" and moderately fine lines, gave results within a very close percentage of the actual surface, but with flat floors and very fine lines the error becomes serious, and as much as 10 per cent. in excess of the truth in extreme cases.

Mumford's Rule suits every class of ship, and the error is seldom over 5 per cent., and generally much less.

Seaton's Rule suits all classes of ship, and the error is generally even less than that of Mumford's when the draught of water is not less than one-third the beam. The examples on page **47** illustrate the above.

Limitation of Speed due to Form and Size.—That each form of ship has a limit to speed is well known; also it is equally obvious that the length and size of a ship must have considerable influence on that limitation. A small ship with a prismatic coefficient of 0·700 could not be driven economically at a higher speed than 10 knots; any increase in power developed is employed chiefly in wave making, whereas a ship 500 feet long of the same form could be driven economically at 19 knots per hour. Again, for 19 knots with a ship 300 feet long the prismatic coefficient should not exceed 0·610, otherwise the power would be excessive by comparison.

In making investigations on this subject the author found that in this case also $\sqrt[4]{L}$ was one of the leading factors for determining these questions, and the formulæ he arrived at involving it he has found to give results in agreement with good practice; they are as follows:—

(1) The highest prismatic coefficient for a speed S and a length L is

$$f = 0.39 \sqrt[4]{L} \div \sqrt[3]{S}.$$

(2) The fastest economic speed for a length L and a coefficient f is

$$S = \left(\frac{0.39 \sqrt[4]{L}}{f}\right)^3.$$

(3) The shortest length of ship having a prismatic coefficient f for a speed S is

$$L = \left(\frac{\sqrt[3]{S} \times f}{0.39}\right)^4.$$

Froude's investigations on skin resistance led him to fix a set of values for j varying from ·00963 for ships 50 feet long to ·0088 for those 500 feet long when the index value for v is taken at 1·825. Some more recent investigations have shown that that index value requires modification, and that really—

Index value of n for ships from 100 to 500 feet long is 1·829 for modern enamel paints when quite clean, and 1·827 for clean bright copper and zinc, and 1·843 for these metals when rough from corrosion.

RESISTANCE OF SHIPS. 47

TABLE IV.—EXAMPLES OF STEAMSHIPS SHOWING THE WETTED SKIN AS ASCERTAINED BY THE SEVERAL FORMULÆ.

Name of Ship.	Dimensions.			Displacement.			Immersed Mid-section.	Wetted Skin as by				K.
	Length.	Beam.	Mean Draught.	Tons.	Block Coefficient.	Prismatic Coefficient.		Kirk.	Mumford.	Seaton.	Measurement.	
T.S.S. "Oceanic,"	685	68·5	32·5	28,500	0·656	0·691	2,113	75,365	68,500	69,095	68,830	10·0
T.S.S. "Kaiser Wilhelm II."	680	72·0	29·5	26,560	0·644	...	...	...	64,899	65,341	66,040	9·4
T.S.S. "Deutschland"	662	67·3	28·8	23,900	0·631	...	...	...	60,520	60,236	59,200	9·7
T.S.S. "Armdale Castle"	570	64·2	30·0	21,734	0·700	...	...	...	53,910	56,400	55,500	8·9
T.S.S. "Empress of Britain"	550	65·5	30·0	21,233	0·701	...	...	...	52,580	54,753	54,500	8·4
T.S.S. "Campania"	600	65·3	25·0	18,000	0·644	0·682	1,540	51,440	50,290	50,290	49,620	9·2
T.S.S. "Durham Castle"	475	56·5	28·0	15,973	0·764	...	...	...	42,750	45,180	43,850	8·4
T.S.S. "Savoie"	557	60·0	25·5	15,400	0·634	...	...	...	45,334	45,300	41,950	9·3
T.S.S. "Fürst Bismark"	470	55·0	26·0	14,007	0·753	...	...	...	39,762	41,657	39,850	8·6
U.S.N. "Alabama"	368	72·0	24·0	11,734	0·644	...	...	...	32,079	32,508	32,460	5·1
T.S.S. "Pennsylvania"	430	50·0	17·5	8,926	0·830	...	...	...	30,638	30,910	30,520	8·6
S.S. "Britannic"	450	45·2	23·6	8,500	0·620	0·714	926	33,943	30,664	31,913	32,578	9·9
H.M.S. "Alexandra"	325	63·7	26·1	9,432	0·612	0·725	1,405	29,888	27,183	28,690	29,620	5·1
H.M.S. "Sultan"	325	59·0	24·9	8,714	0·639	0·711	1,320	29,847	26,065	27,563	28,140	5·5
H.M.S. "Captain"	320	53·2	24·8	7,672	0·626	0·721	1,176	26,967	24,160	25,830	26,220	6·0
H.M.S. "Temeraire"	285	62·0	27·0	8,571	0·629	0·723	1,455	26,948	24,225	25,770	26,900	4·6
H.M.S. "Swiftsure"	280	55·0	24·8	6,537	0·600	0·715	1,140	23,481	21,045	22,060	22,120	5·1
H.M.S. "Vanguard"	280	54·0	22·5	6,076	0·626	0·705	1,077	22,419	20,160	21,145	21,750	5·2
S.S. "Assiniboia"	336	43·5	18·0	5,270	0·703	...	...	...	20,160	21,200	20,400	7·7
T.S.S. "Princess Victoria Louise"	400	47·0	16·5	5,650	0·638	...	...	...	22,889	22,824	23,400	8·5
T.S.S. "Princess Charlotte"	330	46·5	14·0	3,357	0·550	...	...	...	16,000	15,456	15,650	7·1
T.S. "Lady Fraser"	270	38·0	14·0	2,160	0·527	...	...	...	11,610	11,530	11,446	7·1
T.S.S. "Viper"	315	39·5	12·0	2,161	0·532	...	...	...	12,965	11,800	12,750	8·0
T.S. "Hazel"	260	36·0	12·0	1,754	0·549	...	...	...	10,220	10,020	10,240	7·3

As it is in a general way more convenient to deal with these questions on a reference value per square foot at 10 knots speed, the following table gives the resistances per square foot of surface of the different materials as probably actually found in practice as deduced from the foregoing experiments and observations on ships on trial.

TABLE V.—AVERAGE RESISTANCE PER SQUARE FOOT OF DIFFERENT MATERIAL AND CONDITIONS IN ACTUAL PRACTICE.

	Under 100 Feet Long.	Under 200 Feet Long.	Under 400 Feet Long.	Under 600 Feet Long.
(1) Copper bottom, new, bright and clean	·898	·874	·856	·850
(2) ,, ,, clean	·966	·943	·926	·926
(3) ,, ,, corroded	1·250	1·230	1·175	1·140
(4) Enamel paints, best kinds freshly done	·970	·944	·916	·900
(5) ,, ,, good and clean	1·030	1·000	·970	·950
(6) Ordinary paint; tar, etc.	1·140	1·100	1·070	1·050
(7) Tar and plumbago polished	·922	·900	·884	·875
(8) Fine grass on paint	5·200	4·800	4·670	4·540

The resistance or losses due to wave-making and eddying are called generally *residual losses*, and may be calculated with a fair amount of exactness by the following rule :—

$$\text{Residual resistances} = x \, \frac{M \times B \times v}{\sqrt{B + zL^2}}.$$

M is the area of immersed midship section in square feet.
B is the extreme breadth in feet.
v is the velocity in feet per second.
K is the ratio of length to breadth; that is, $L \div B$.
H is the ratio of the length to velocity square; that is, $L \div v^2$.
x is a factor got by $150 \div K^2$.
z is a factor got by $0·55 \div \sqrt[3]{100H} = \dfrac{0·118}{\sqrt[3]{H}}$.

(*a*) The total resistance is then made up of the skin resistance *plus* residual resistance.

It was the common practice formerly to determine the total resistance by the following formula, where D is the displacement in tons and S the speed of the ship in knots.

RESISTANCE OF SHIPS. 49

(b) Total resistance $= D^{2/3} \times S^2$ in pounds.
It was due to the following, viz.:—

$$\text{Resistance} \times S \times \frac{6080}{60} = \text{E.H.P.} \times 33,000,$$

or $R \times S = \text{E.H.P.} \times 325$ foot-lbs.

E.H.P. $= k \times$ I.H.P. Then $R \times S = $ I.H.P. $\times 325k$,

$$\text{and I.H.P.} = \frac{R \times S}{325k}.$$

By the Admiralty speed formula, I.H.P. $= \dfrac{D^{2/3} \times S^3}{C}$.

By substituting the above value of I.H.P. the following holds:—

$$\frac{R \times S}{325k} = \frac{D^{2/3} \times S^3}{C}; \text{ or } R = \frac{D^{2/3} \times S^2}{C} \times 325k.$$

Now if the efficiency, etc., of the ship is high, k will be high, as will also the value of C. Taking k at 0·6, then

$$R = \frac{D^{2/3} \times S^2}{C} \times 195.$$

Then if $C = 195$, which would be its value with the old ships of Rankine's day, $R = D^{2/3} \times S^2$.

To-day we have ships with an efficiency as high as 0·70; the coefficient C, then, would be 227·5. If the efficiency is as high as 0·75, then C would be 244.

Test of efficiency of the propulsive powers of a ship can be, and has been, gauged for more than fifty years by the magnitude of this coefficient; as also by that of another coefficient which takes into account the size of the immersed midship section. The formula is:—

C = area of immersed midship section × speed ÷ I.H.P., usually written $\dfrac{A \times S^3}{\text{I.H.P.}}$

The older engineers set more value on this latter criterion than on the former. To-day, however, the latter is not often referred to.

Example.—To find the wetted skin resistance and E.H.P. of a steamer 300 feet long, 38 feet beam, and 13·5 feet mean draught. Her displacement is 2400 tons, prismatic coefficient is 0·61, block, coefficient ·546, and she is to steam at 18 knots. Allowance for keel 6 inches.

4

Wetted skin by Mumford's rule $= 300\{1.7 \times 13 + 0.546 \times 38\}$
$= 12,854$ square feet.

,,   ,,   Seaton's formula $= 70.4 \times 2400\%$
$= 12,622$ square feet.

Take the wetted skin as 12,800 square feet and allow 1·05 as the resistance per foot at 10 knots. Then

$$\text{Skin resistance} = 12,800 \times 1.05 \times \left(\frac{18}{10}\right)^2 = 43,545.6 \text{ lbs.}$$

$$\text{The residual resistances} = x \frac{M \times B \times v}{\sqrt{B + zL^2}}.$$

Here $\quad v = 30.4; \quad\quad x = 150 \div 63$
$M = 460$ square feet; $z = 0.118 \div \sqrt[3]{0.333}$.

Substituting these values

Residual resistance $= 10,242$ lbs.
Total resistance $\quad = 43,520 + 10242 = 53,762$ lbs.
Resistance H.P. $\quad = \dfrac{53,762}{33,000} \times \dfrac{18 \times 6080}{60} = 2971.$

Taking efficiency of ship and machinery at 0·6, then

$$\text{Gross I.H.P.} = 2971 \div 0.6 = 4952.$$

Every steamship is subject to other sources of resistance beyond those already named, so that the gross resistance when steaming with its own machinery is higher than that experienced when the bare hull is being towed or moved by other extraneous means in smooth still water. Of these the following are chief:—

Augmented resistance due to the action of the propeller is a most serious loss with some steamships, while with others it is of little moment. In the case of side wheel steamers the velocity of the stream from and in wake of the wheels is higher with respect to the ship than is the "still water," consequently the friction of the skin from the wheels to the stern is higher than that of the towed ship; but this increase is comparatively small. With a stern paddle wheel and a screw propeller there is the feed or stream flowing to the propeller acting on the ship and increasing the friction for some feet before the stern; but the loss from this source is trifling compared with that due to the loss of "head," due to the rapidity with which the water is withdrawn from the stern of the ship. In fact, with a large screw of fine pitch the reduction in pressure on the stern may amount to nearly that due to the removal of an area equal to that of the screw disc, especially when the ship has a full after-body.

Air resistance was another fruitful scource of loss of efficiency in the days when ships were fully masted and sparred as was a sailing ship. High superstructures also form great obstructions to easy passage through the air. To-day steamships have only apologies for masts and no other spars of consequence, but many of them have huge and numerous superstructures, and ocean passenger steamers have row upon row of decks more or less broken so as to cause great obstruction, and then deckhouses and erections above them, so that the loss due to air resistance is nearly as large as was that of the old frigates and line of battle ships. With a ship moving through still air at twenty-five miles an hour the "head" resistance is considerable, amounting to 3 lbs. per square foot. In addition to this, however, there is the friction of the sides and the thousand-and-one small projections and recesses, all of which will sum up to more than the direct head resistance in some ships.

Depth of water has a great influence on the speed of a ship, for, as a rule, as the ship's bottom approaches the sea or river bottom the resistance increases considerably, and when they are so near that sailors say they are "sucking," that is, there is not sufficient inflow between them to properly support the ship, the resistance is extreme.

On the other hand, there have been instances with small fast craft, as torpedo boats, when a distinct gain in speed has been manifested on coming into shoal water. Attention to this phenomenon has been called by several engineers at the meetings of the Institution of Naval Architects, in whose Transactions may be read some interesting papers and equally interesting discussions on the influence of depths of water on a ship's speed. It would be out of place to do more than call attention to them here, as it is presumed that all trials bearing on the subject of propeller efficiency are made in sufficiently deep water, in fact, the deeper the better.

Smooth water is essential where the highest efficiency of a ship and propeller is desired in all steamers, but especially is it so with small craft which are sensitive to even a ripple. In a general way, however, small waves do not affect sea-going craft to an appreciable extent, while a "billow," even when slight, does so, inasmuch as the ship is then travelling on the arcs of curves instead of on the chords. The larger the ship the larger may be the waves without affecting the resistance or propeller action. If, however, the ship be running parallel to the waves and rolling be set up, the efficiency must be lowered. Also if the almost imperceptible billow has a period synchronising with that of the ship, heavy rolling with consequent loss of efficiency will follow. As a rule, therefore, when accurate results are required from a steamship trial, it should take place in smooth water in a place sheltered from the wind and where there is little or no tidal or other current.

Tidal currents sometimes cause considerable errors in the estimate of speed on measured miles, inasmuch as the flow is never at a constant rate nor steady in direction. The measured mile at Skelmorlie, Wemyss Bay, has the qualities as near to perfection as possible, and so enables the trials of the largest and smallest ships to be carried out and concluded with the least amount of disturbance; consequently the results thus obtained can generally be relied on as being quite accurate.

Trim of the ship affects the speed considerably, especially that of the modern flat-bottom steamer, which, when trimmed by the stern, presents a huge plane at a considerable angle from the line of motion. Ship captains and sailors generally will insist on trimming their ships by the stern, alleging that they not only steer better, but steam faster.

The steering may be better as more rudder comes into action and the centre of longitudinal resistance is nearer the rudder; but the steaming cannot be so good as with the ship on the even keel or on the trim as designed. In old days it was common experience to find a paddle steamer doing better when trimmed by the bow; and it is quite possible for a screw steamer to have less resistance when her bow is depressed so as to free the bottom from direct head pressure. The immersion of the screw must of course be complete, and therefore trimming by the stern when in ballast or half loaded may be excusable. When fully laden, however, so that the screw is well under water, an even keel is the best trim for good results in smooth water, and not bad ones will be obtained in a sea-way.

But notwithstanding all that has been written in text-books and recorded in scientific institution Transactions, in spite of all the investigations made by highly trained intellects both as to the theory and practice of ship design, it is admitted that, to know beforehand the *exact* total of the resistance of a ship, resort must be had to experiment; which, thanks to Dr W. Froude's patient and prolonged investigation, may be made on a model of comparatively small size—

so small indeed that the experiments may be carried out in a tank within a building with appliances suitable for the purpose so delicately constructed and finely calibrated that from the registrations made by them from such models, the needs of the largest ship can be arrived at with accuracy. It need hardly be said, however, that such means as these are costly, and only a few can afford such a luxury. Thanks, however, to the generosity of Mr A. F. Yarrow, such a tank, etc., will be soon at the service of anyone, and may be used for the public benefit by a staff of highly-trained men. Still, seeing that the commercial engineer, who has to give guarantees and suffer penalties, cannot afford to run risks, he must trust to the experimental data to guide him or, as he has generally done in the past, work with a margin of power which thereby insures him. In the past, however, this was a somewhat risky process at times, as attempts were made to drive ships at speeds for which their shape was not suitable; to-day it is easy to determine the limits for each model of an ordinary ship, and by modern sound formulæ to calculate the I.H.P., thrust, etc., with sufficient precision to suit all needs. But, while not being desirous of depreciating the value of tank experiments, it must not be forgotten that more than one failure in the shipping world has been made in spite of the tank guidance.

Variation in total resistance R in destroyers, also the resistances other than those of friction R, are given by Sir William White as follows:—

Up to 11 knots R varies nearly as S^2 and R_1 varies as S^2 at 11·0.
,, 16 ,, ,, ,, S^3 ,, ,, S^4 ,, 14·5.
,, 18–20 ,, ,, ,, $S^{3·3}$,, ,, S^5 ,, 18·0.
,, 22 ,, ,, ,, $S^{2·7}$,, ,,
,, 25 ,, ,, ,, S^2 ,, ,, S^2 ,, 24·0.
,, 25–30 ,, ,, ,, $S^{1·83}$

The following was also observed, that the relation of frictional resistance to total was:—

At 12 knots 80 per cent. in destroyers, 90 per cent. in cruisers.
,, 16 ,, 70 ,, ,, 85 ,, ,,
,, 20 ,, 50 ,, ,, 80 ,, ,,
,, 23 ,, ,, ,, ,, 71 ,, ,,
,, 30 ,, 45 ,, ,,

CHAPTER IV.

ON SLIP—APPARENT, REAL, POSITIVE AND NEGATIVE. CAVITATION. RACING.

THE preconceived idea in early days was that a paddle wheel acted on the water like a pinion on a rack, and a screw propeller as a screw bolt in a nut, or as a screw pile in a stiff clay bank. The difference between the theoretical and the actual advance of the ship on these suppositions was due to the slipping of the propeller through the semi-resisting medium, viz. the water acting on it; it was therefore spoken of as the slip of the *propeller*, whereas it is really the slipping of the *water* with respect to the still water.

In the case of a paddle wheel, even, it is not easy or practicable to determine what is its exact mean velocity as a propeller, and much less can the actual mean effective velocity of the stream from it be obtained even when the ship is running in still, smooth water.

With a radial wheel it was customary to make all calculations by taking the mean radius of action as the centre of the float; but it is obvious that when the float is considerably submerged its centre will not coincide with the middle of the stream produced by it. With the feathering wheel the action was assumed to be at the axis about which the float swung in feathering—that is, at the centre of the suspension pins or gudgeons. When these were in the middle of the float, then the rule coincided with that for radial floats; but it often happened, and it is still not uncommon to find, that the gudgeons are at the centre of pressure of the float when fully immersed and not at its middle line. In neither condition, however, can it be proved that the velocity thus calculated is the real mean velocity of the stream produced by the floats.

In the case of the screw propeller the difficulty of ascertaining the actual mean velocity of the stream set in motion by it is still greater, even when the screw is a true one and the form of blade a simple geometrical figure. With a variable pitch and unsymmetrical shape of blade it is impossible to do so with any degree of accuracy. In practice the speed of a screw is calculated by multiplying the revolutions per minute by the pitch; in the case of a varying pitch it is a common method to take a mean by adding the extremes and dividing by two; it need hardly be said, however, that this is a very unsatisfactory method, although perhaps a convenient one. The old Admiralty method is, on the whole, more satisfactory, viz. to take and state the velocities due to the extreme pitches and ignore a mean quantity throughout.

The hydraulic propeller being enclosed, and taking as well as discharging its water through tubes or enclosed channels, the mean velocity of flow can be easily ascertained, and also where the centre of force is situated; besides which, the quantity of water which passes through the propeller in a given time can be calculated correctly; whereas with the paddle wheel and screw, not only cannot the mean velocity of the stream be either calculated or measured, but the quantity of water projected by them is unknown in actual practice.

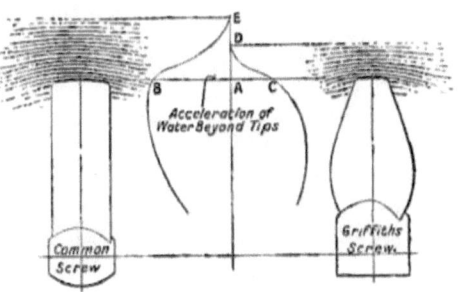

FIG. 13.—Showing the Influence of Screw on the Water beyond the Tips.

It was often assumed that the sectional area of the race from each wheel was equal to the length of float multiplied by the immersion of its lower edge, providing the wheel was not "drowned" by too much immersion. Taken altogether, the stream cannot be more than this when the immersion of the upper edge is not less than the breadth of the float; it will be safe to take the area of each wheel stream as not more, in any case, than twice that of one float.

As the top edge of the float of a feathering wheel was seldom immersed to the extent of the radial float, and as there were only about half the number of floats on it, it is sufficient, nowadays, to assume the race's section as equal to the area of one float.

The stream section from a screw propeller can only be assumed to equal the area of the circle made by its blade tips less that shut off by its boss, and it is customary to take this for calculations and to suppose the tail race to be a hollow column. There will be, however, great differences in the rate of flow from that at the circumference and those streams nearer the centre. It is more than probable that there is a cone of dead water about the axis of every screw race, and hence a hollow column or cylinder with a velocity of flow varying from zero at the middle to the maximum near the circumference. The diameter of the column of water actually in motion near the screw must be in excess of that of the screw; especially will this be so with broad-tipped screws running at high revolutions; for the friction of the water set in motion axially by the portion of blade near the tips will be sufficient to induce a stream beyond it to flow with, although of course gradually lagging behind, it as the distance from the screw increases (see fig. 13). So also in the case of the paddle wheel, the natural stream or race made by the floats will induce motion gradually in the water in advance of the wheel, as well as to start into motion the water through which it passes as it goes through the wheel (fig. 14). That is, the actual race extends beyond the float ends as it does beyond the blade tips of a screw.

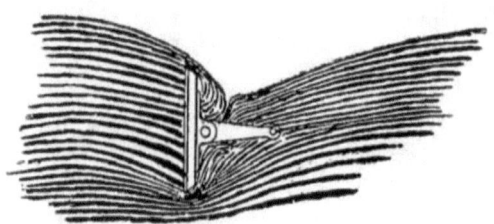

Fig. 14.—Flow of Water to a Paddle Float.

Speed of paddle wheel, whose diameter in feet D is taken at the centres of floats in the case of floats of the radial wheel, and at the float pins of feathering floats; in the case of those considerably immersed radial floats D should be taken at the inner edges. The revolutions per minute R; then

$$\text{Speed per minute} = \pi D \times R \text{ feet.}$$
$$\text{,,\ \ ,, hour} = \frac{\pi D \times R \times 60}{6080} = \frac{D \times R}{32 \cdot 25} \text{ knots.}$$
$$\text{,,\ \ ,, second} = \frac{D \times R}{19 \cdot 1} \text{ feet.}$$

Speed of screw whose pitch in feet is P and revolutions per minute R.

$$\text{Speed per minute} = P \times R \text{ feet.}$$
$$\text{,,\ \ ,, hour} = \frac{P \times R \times 60}{6080} = \frac{P \times R}{101 \cdot 3} \text{ knots.}$$

SLIP; CAVITATION AND RACING.

Speed of ship being taken as v; speed of propeller as V in feet per second.

Apparent slip of propeller is $V - v$; and

Real slip is the velocity imparted to water stream by the propeller.

Slip of propeller per cent. $= \dfrac{V - v}{V} \times 100$.

It is in this form of a percentage of the speed of the propeller that engineers usually express its performance.

In the early days of marine engineering the efficiency of a propeller was expressed by $\dfrac{v}{V}$, so that perfect efficiency would be when this has the value unity; this was only possible when the speed of the ship equalled that of the propeller—that is, when there was no slip.

Real slip, however, is the most important thing to determine, for it is the mean velocity actually imparted to the water by the propeller compared with that of the water unaffected by the ship or her propeller, and commonly called "still water." It has been stated that in practice this cannot be calculated with even moderate accuracy, and only in the case of the hydraulic propeller can it be ascertained.

Thrust.—The forward motion of the ship is due to the reaction produced by projecting backwards a heavy body whose mass is M at a velocity $V - v$. This body may be water, or it may be cast iron or the rack or "a nut"; for if the rack were heavy and free to slide without friction there could be still a pressure at its teeth due to its inertia when the pinion was turned, which would be transmitted to the pinion axle as thrust, just as it is in a lathe saddle with the rack secured and prevented from slipping.

Now, theoretically the quantity of water operated on by any propeller is measured by multiplying the section of the tail race or stream issuing from it in square feet, by the mean velocity of flow, say in feet per second.

Taking A as the area of section and (assuming the real slip to be equal to the apparent) V as the mean speed of flow in feet per second, 64 lbs. as the weight of a cubic foot of sea water, and gravity as 32,

(a) Volume of water per second $= A \times V$.
(b) Weight   ,,   ,, $= A \times V \times 64$.
(c) Mass   ,,   ,, $= \dfrac{A \times V \times 64}{32} = 2A \times V$.

The acceleration or velocity imparted to it is $V - v$ per second.

(d) Then momentum $= 2A \times V (V - v)$ lbs.

This is the measure of the propelling force, and the reaction is called *the thrust*.

If a ship is at rest with respect to the water in which it floats and is to be set in motion by the thrust or reaction of the water stream projected from it, the ship's inertia must first be overcome; and until that is accomplished the flow of the stream will be at the same velocity with respect to the water as to the ship; then as v is 0 the slip per cent. $= \dfrac{V-v}{V} \times 100 = 100$. If the stream issues at a constant velocity from the ship its velocity with respect to the water will decrease as the ship's velocity increases; moreover, this velocity of the ship will continue to increase, as will also its resistance, the latter varying as the square of the speed through the water.

When the whole propelling force is taken up with, and so equals, the ship's resistance, there can be no further increment of speed.

If that force or thrust be multiplied by the space moved through in feet per minute by the ship, it will give in foot-pounds the work done in propelling the ship; and if this be divided by 33,000, it gives the net Effective Horse Power (say E.H.P.) required to drive the ship at that speed.

The efficiency of engines and propellers together is therefore E.H.P. ÷ I.H.P.

Hence $E.H.P = \dfrac{A \times V (V-v)}{1800} \times \dfrac{v}{33,000} = \dfrac{A \times V \times v (V-v)}{59,400,000}$.

Rankine's formula for thrust in pounds deduced from this, and where S is the speed of the propeller and s that of the ship in knots, is as follows, viz:

Thrust $= A \times S (S - s) \div 5.66$ for sea water and 5.5 for fresh water.

It is manifest that if motion is imparted to the water by a propeller, the stream of water must issue from it at a higher velocity than that at which it entered; that is, it is higher than the velocity of the ship; there must therefore be always positive *real slip*. Had not experience demonstrated the contrary, it would have been insisted on that there must also always be positive *apparent slip*. Probably even the *possibility* of no apparent slip would have been denied. The surprise of the pioneers of screw propulsion therefore may be imagined when they discovered that certain screws possessed the virtue (?) of driving a ship faster than that due to their pitch. Certainly a virtue, seeing that in their eyes slip was an evil thing and an unnecessary loss. It must, however, have been a puzzle to them as well as a mortification to find negative slip showing itself, since they measured the efficiency of a screw by dividing V by v, and efficiency could not possibly be greater than 1.

To this day negative slip remains still as something of a mystery, in spite of the assurances that it was and is still due to the effect of wake currents of the ship, to the flowing of water in behind the screw when the fullness of the ship's stern lines has prevented the screw race from otherwise "running full." As the phenomenon is still occasionally met with, it is worth while making further investigations of the facts to discover the cause or causes of it.

Negative apparent slip might be an imaginary or unreal thing arising from faulty observation and inaccurate measurements.

For instance, when a screw is simply cast and is rough from the sand, it is seldom shaped so that the pitch throughout is exactly that designed; it will differ in different blades and be different in parts of the same blade. If such a screw is simply trimmed up and even *carefully measured*, no one can say what is really its *effective* mean pitch. But in early days, before machines were employed as they are to-day in reducing a rough casting to an accurate portion of mechanism finished to the designed dimensions, the foundry man was accustomed to take greater care in moulding so that as little trimming of the casting as possible had to be done. His propellers were fairly accurate as to pitch. It may be taken therefore that apparent slip occurred, and may occur even with propellers properly shaped and as carefully measured.

In the case, however, of Woodcroft's and other propellers designed with a varying pitch, there is better reason for rejecting the suggestion that any mean pitch calculated from the measurements could be of any value for scientific investigations than for questioning the accuracy of the measurements. Such propellers as these, as used at one time in the Navy, frequently, but not always, displayed the tendency to negative apparent slip. Strange to say, too, the percentage of slip at low speed was much higher than at full speed, although the wake disturbance would then, of course, be less active.

Another source of error which might have accounted for this extraordinary phenomenon was the rough and ready way in which trial trips were carried out then as compared with the care that generally obtains now. For example, measured miles were always in tidal waters more or less exposed, and at some the stream would run strong and vary considerably; accuracy of speed of ship is impossible of estimation under such circumstances. But negative apparent slip was as observable at Stokes Bay as at the Maplins, and to-day it occurs sometimes even at Skelmorlie. Moreover, in fact, so far as the disturbing causes of the old measured miles are concerned, one would rather expect increase in positive than development of negative slip.

The counting of revolutions was often done by merely taking account for only one minute or even less, with a common watch, and by assuming that the rate of revolution was constant throughout the mile. The tendency, however—and it certainly was the temptation in those days—was to state the revolutions higher rather than lower than those actually made, seeing that the engine contractor had always to guarantee I.H.P., and very rarely was responsible for the speed. But in many of these trials it is evident the revolutions must have been carefully counted.

It may be suggested that, inasmuch as all screw propellers might, with even comparatively small differences in pressure per square inch on the leading and following portions of the blade, twist out of pitch, these old broad tipped blades with comparatively small bases or roots would be peculiarly susceptible to such changes. Further, that with the pressure on and normal to the blades tending to bend them towards the bow, such a displacement in that direction might throw the screw out of pitch. In some cases, perhaps, the real slip may differ from the apparent by a larger amount than it otherwise would be from such causes, especially when the blades are of bronze and thin.

All these possible causes were, however, examined and discounted years ago, and the consensus of opinion settled the matter in the way above indicated. If, however, wake currents caused by the ship were the cause—and no one will doubt that they are most disturbing elements in their effects on screws—how did it come about that in the same ship two screws having the same diameter should in one case run with positive slip and in the other with negative? Further, what was there to make wake currents so very pernicious from 1864 to 1870? for although negative apparent slip had been observed as early as 1841 to a small extent in H.M.S. "Rattler," and in 1853 on H.M.S. "Miranda" when on her trial trip, it was 7·85 per cent. slip; it was not, however, of common occurrence till 1864, when H.M.S. "Achilles," a long and fine-lined ship, showed as much as 13·0 per cent. negative slip in Stokes Bay; a similar but somewhat longer ship, H.M.S. "Agincourt," showed 17·14 per cent. negative at Plymouth in 1865; and in the same year H.M.S. "Favourite," "Bellerophon," and "Amazon" showed the same phenomenon, together with poor speed coefficients. In the following years other ships developed the same objectionable features. Finally, in 1877 H.M.S. "Iris" was such an abject failure as regards speed for power that a more careful investigation was made in her case than in the others.

It is noticeable throughout that in every case the propeller was four-bladed, and generally with a varying pitch. But plenty of four-bladed screws had run in the mercantile marine, and not a few in H.M. service, successfully, or at any rate without *negative* slip; and since then thousands of four-bladed screws have done the same. Those, however, with which the negative slip occurred always had blades with broad tips and were of large diameter, and consequently had large area of acting surface; their pitch ratio was always small, often less than one. In the case of H.M.S. "Amazon," "Bellerophon," "Lord Clyde," "Pallas," "Zealous," and others, such screws were replaced by two-bladed ones of the Griffiths design, with the result in every case of the apparent slip becoming positive and appreciable; and while in one case the speed was only the same as with the old screw, in the others it was appreciably higher.

In the case of most of these ships the diameter of the new screws was the same as that of the old; why, then, were they not affected by the wake currents so as to show negative slip? It can hardly have been by reason of the two blades.

In the case of the "Iris," be it remembered, she was a twin-screw ship and her propeller scarcely in the wake, seeing she was an exceptionally fine-lined ship. Her original screws were like those of the other ships above named, viz. of large diameter with broad tips and a varying pitch, and ran with negative slip; as indeed they did when two blades from each was removed and the surface consequently reduced by 50 per cent. With true screws of Griffiths type of practically the same diameter, with two blades, the speed was very much higher and the slip over 5 per cent. positive, while with four-bladed screws of leaf shape but with 2 feet less diameter, the same high speed was obtained as with the Griffiths, and with positive apparent slip, 3 per cent.

It seems from the above that the wake current theory does not account for the whole, if for any, of the phenomena; but it would appear as if the greater part, and perhaps the whole, was produced by the screw itself.

In the first place, it may be questioned whether any propeller having a considerable thickness of blade with cross sections approximately segments of circles, can act in the same way as would one whose blades are thin but rigid and of uniform thickness; especially is it open to doubt if the effective pitch even approximates to that of its acting face or designed pitch. Next, that with a pitch increasing from leading to following edge of blade, the actual acting surface will probably be a water one whose angle is greater than that portion of the blade with maximum pitch. If so, the pitch from which to deduce the slip is unknown, but in any case it must be greater than the greatest pitch. (See fig. 15.)

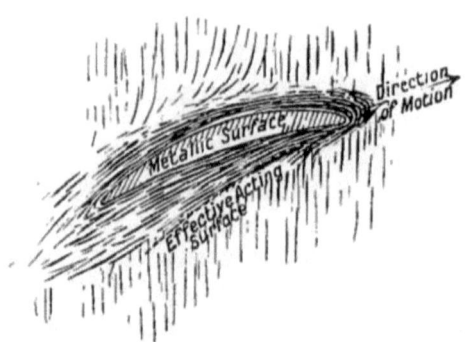

Fig. 15.—Passage of a Curved Screw Blade section through the water.

Further, the stream of water flowing into a screw may vary in cross section and consequently in velocity.

Suppose a ship to have a screw propeller cased in as shown in fig. 16 so that it is supplied with water at the orifice in sufficient quantity at a velocity v, which is that of the ship through still water. If the propeller does not revolve but the ship continues its course by sail power, or on being towed, water will enter D at velocity v and

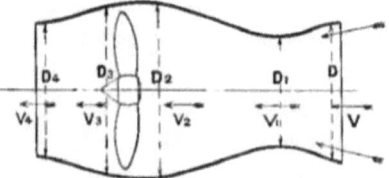

Fig. 16.—Showing a Screw working with Negative Apparent Slip.

leave D_4 at the velocity of $v_4 = \left(\dfrac{D}{D_4}\right)^2 \times v$. Of course this is assuming there is no resistance, frictional or otherwise, in the passage. If D_4 is larger than D, then the flow will be less than v, and consequently the water will be carried on with the ship with a loss measured by $v - v_4$.

If the screw is set in motion there will be a suction of water and

SLIP; CAVITATION AND RACING.

an acceleration given to it so that its velocity of flow is increased beyond v and is now v_1; to keep this up through D_1 the orifice D must be larger than D_1 or the channel must be reduced so that the flow at D is simply that due to the ship's motion v.

If, then, D is less than D_2 the flow of the stream will be less than v; and further, if v_3 is less than v the speed of the propeller, or $p \times r$, is less than v.

Consequently, taking slip by the usual criterion, viz., $v_3 - v$, it will, under the foregoing circumstances, be a negative quantity.

If D_4 is equal to D, the velocity of discharge will be the same as that of the ship, and water will be delivered "at rest" with respect to still water.

If D_4 is contracted, as more than one inventor of an hydraulic propeller has suggested, the velocity of tail race will be greater than that of the ship, and the water will be projected through the still water, with much consequent waste of energy.

In all screw steamships there is a flow at the stern something like that through the channel of fig. 17. In bluff-cargo steamers, as also in the old sailing war ships, the

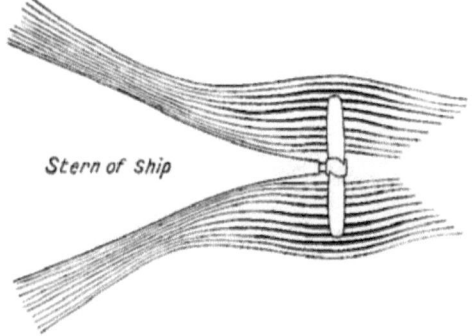

FIG. 17.— Flow of Water from Stern to Screw.

"run" was comparatively short and curved at the water line, even more than as shown in fig. 17, so as to give a good flotation plane for stability. The feed to the propeller causes an increased flow before reaching the screw, as well as a raised wave, to permit of the quantity of water passing into the race. In front of the screw it spreads out by diversion, and at the screw the supply to the tips is assured by this dispersal. Behind the screw, contraction takes place, due to the fall in velocity at the other part of the race, and the rope-spinning action of the screw ensues which Fitzgerald noted.

Reduce the tips either by reducing the diameter or by narrowing them as Griffiths did and the tendency of the water to flow out to give them the necessary supply is reduced; D_2 is virtually much smaller, and so v_2 and v_3 become greater than v, consequently $v_3 - v$ is then a positive quantity.

Example.—A screw 10 feet diameter and 10 feet pitch turns in a casing 11 feet diameter and is supplied with water through an orifice at the bow 9 feet diameter. The ship travels at ten knots and the *real* slip to do this is 10 per cent. What will be the apparent slip of the screw?

Here v is 16·9 and V 18·8 per second.

$$R \times P = 16·9 \times \frac{9^2}{11^2} = 11·3 \text{ feet per second.}$$

Revolutions will therefore be $= \dfrac{11·3 \times 60}{12} = 56·5$ per minute.

The real slip is $V - v = 18·8 - 16·9$ feet per second, or 10 per cent.

The apparent slip is $V_1 - v = 11·3 - 16·9 = -5·6$ feet, or 49 per cent. negative.

Now when a screw is moving through the water its tips are of course following a spiral path whose diameter is that of the screw, and its length the distance passed through by the ship. Consequent on its roughness and friction it will drag with it a certain amount of water, which water will set up currents in that next it, and so on until the action is too feeble to induce farther water movement. With screw blades having broad tips the drag will be more than that set up by a Griffiths or leaf-shaped blade. A screw, then, may be said to work in a hollow; that is, it sets up a current in the midst of still water and works as if cased, as in fig. 17; the stream lines of the ship induce a flow of water to the screw in a direction much as shown in that illustration; it then spreads out as it approaches the screw to supply the demand of the blades, which, if broad at the tips, is greater there than if the tips were narrower. If a propeller of smaller diameter is fitted, the water does not spread out so much and consequently does not lose its velocity to the same extent. H.M.S. "Archer" was fitted with a two-bladed screw 12·5 feet, and only 7·7 feet pitch, with the result that the apparent slip was negative and as much as 17·68 per cent. The diameter was cut down by degrees, when the negative slip decreased; and when the diameter was 10 feet, it changed to 0·45 per cent. positive, and a further reduction to 9 feet produced 7·24 per cent. positive, although the pitch ratio was then only 0·86.

Influence of Pitch Ratio on Apparent Slip.—It is seen by experiments with the "Archer" that by reducing the diameter of the screw foot by foot, the apparent slip was gradually changed from being largely negative to decidedly positive. In her case, however the pitch ratio was always less than unity.

From 1843, when the screw was first introduced into the Navy, till 1870, the records show that fifty-six ships were fitted with screws the pitch ratios of which did not exceed 1·05. Of these, twenty-one, or 37·5 per cent., developed apparent slip of negative value, while ten developed positive slip exceeding 15 per cent. Examination shows that all these latter were converted sailing ships having very full sterns into the deadwood of which the screws were fitted, and that their speed was from 6·53 to 9·5 knots, so that their screws were comparatively small and ineffective. Neglecting them, the number under review is forty-six, and of them those showing negative slip were 45·6 per cent.

It is also remarkable that, as a rule, the smaller the pitch ratio the greater was the percentage of negative slip and the less the speed coefficients. For example, the "Amazon" screw had a pitch ratio of 0·76 with a negative slip of 23·8 per cent., and $\frac{D^2 \times S^3}{I.H.P.} = 117$, while the "Simoom," with pitch ratio 1·03, the negative slip was only 2·36 per cent., and the speed coefficient 241.

But inasmuch as two factors come into play to make the pitch ratio, the variation may be made by varying either, so that while talking of pitch ratio affecting the question of apparent slip, it may be that it is one of these, and not their combination, which is at fault. It may be, and it seems more than likely, that excessive diameter is the disturbing cause, and the "Archer" experiments might be taken as confirming this view; but when it is seen how often it happens that even the same screw with comparatively small changes of pitch changes the character of the slip, it is open to doubt. Further, it may be noted that in the case of H.M.S. "Iris" the original four-bladed screw showed negative apparent slip at all speeds, but when two blades were removed, while the diameter and pitch remained unaltered, the pitch became slightly positive at all speeds. Also in this ship when fitted with two-bladed screws of the Griffiths type, and even of larger diameter than the original, while the slip was positive at high it fell off to being negative at reduced speeds. Even the four-bladed replace screw of this ship, although reduced to 16·3 feet diameter and 19·96 feet pitch, and consequently of a pitch ratio of 1·205, showed negative slip at slow speeds. All these things point in the direction of attributing negative slip to large diameter; for in the case of the "Iris" even these smaller screws were really too large, for to-day a cruiser whose I.H.P. ÷ revolution is much larger (75) than the same ratio (42) of the "Iris," has screws only 15·75 feet diameter.

Negative slip was a common experience also with tramp steamers of fairly good power, when the screws were almost invariably of large diameter and small pitch ratio, but seldom under unity. It is also not an uncommon thing to find with merchant steamers—especially those built for cargo and passengers, where the lines are only just fine enough for the maximum speed—that whereas the screw (generally a larger one, too) runs at full speed with positive slip from 5 to 7 per cent., at the slow speeds there will be a change to negative slip; this, again, looks as if excessive diameter were the real *fons et origo*.

Cavitation.—When a screw is at work, water may flow to it in other than the usual stream lines mentioned above. For instance, at starting with full steam on the engines the screw may race away at high revolutions before the inertia of the water is sufficiently overcome to flow in a stream to it; and as the flow in any case is due to gravity, the water will flow in from all around. If, however, the action of the screw is so violent that the head of water is insufficient to cause the flow in of water in time, there will be a void place about the screw or a considerable number of small void places, so that the screw is working partly in air and partly in water, and therefore the value of M, and consequently the thrust, is materially reduced. If the action of a screw with the ship under way is very violent, that is, the rate of revolution too high for the immersion of the screw, cavitation will also take place ; so much so that the efficiency of the propeller is seriously impaired. Such a state is indicated by a falling off in the increase of thrust coinciding with an increase in the percentage of slip as the revolutions increase.

A similar result is experienced when a screw has too small a blade surface for the power transmitted to it. In this case there is at first the increase in revolution due to the higher power, and, of course, some increase in thrust until the pressure per square inch on the blade is greater than that due to the head of water. That on the upper blade will become broken up, while that on the lower may be unaffected; the result is a great difference of total pressure on them, which tends to shake the ship, and, as sometimes happens, shake it violently. This is due to the formation of cavities containing air evolved from the sea water. It is also not unlikely that a propeller with too small a surface, especially if of coarse pitch, moves round at times surrounded by the water it has set revolving with it, causing it to act as if it had infinite pitch or none. This is what takes place when the wake column is ruptured, but it can hardly be called cavitation.

Mr Sidney Barnaby was the first to call public attention to the phenomenon now known as *Cavitation*, and while others had observed before that time the variable action and falling off in thrust of screws with insufficient blade area, to him is due the honour of having investigated the causes both mathematically and by experiment, and traced them to their source. He also points out that "if the velocity which has to be imparted to the water in order that it may keep in contact with a portion of the blade situated at a depth h is less than $2gh$, then, even if the blade break the surface, there will be no loss of efficiency." He deduces from this the opinion that to this is due the good results observable in barges on canals working with only partly immersed screws.

This was the idea of David Napier in 1851 and of others since, but their experiments on these lines proved to be failures, although it must be said that in Napier's case the poor results were undoubtedly due to the bluff water lines of the ship employed by him, whereas for such a purpose he should have had one with a very fine run.

The rate of flow of water to a screw wholly submerged, where cavitation is caused by the air abstracted from sea water, will be that due to a "head" equivalent to the atmospheric pressure added to that in the cavities together with the *head* of the water itself; but Mr Barnaby is of opinion that rupture takes place long before the mean pressure on the blade has reached 30 lbs. per square inch.

Since Mr Barnaby read his paper to the Institution of Naval Architects in 1897, cavitation has had a fresh interest added and become of much greater importance, owing to the introduction by Mr Charles Parsons of his steam turbine as a marine motor. High speed of revolution is necessary for him to get decent efficiency of motor, and consequently he began by running screws at rates of revolution undreamed of before. Even now the speed is high in all sizes of ships; it is 190 in the "Lusitania," 500 in large cruisers, and as much as 1200 in small high-speed ships. Such revolutions are only possible with propellers of comparatively small diameter, and they, to avoid cavitation, must have blades of great width and consequently immense surface.[1]

---

[1] Since the "Lusitania" and "Mauretania" have been on service it has been found necessary and advantageous to fit them with new screws having four blades in place of the three as shown in the frontispiece.

Racing of screws takes place when from some cause there is a temporary increase in the rate of revolution. It may be due to cavitation, and often is when the acting surface of the propeller is somewhat small in proportion to the power. This is observable when the ship is running in quite smooth water under conditions of least disturbance; for then, when the sea is quite clear, the propellers may be seen to break up the water as if they were pulverising it instead of boring through it. That propellers with very narrow rounded tips tend to do so is shown by the stream of bubbles flowing now and again from the tips although well submerged.

When the propeller is only barely submerged the surface is broken through by the presence of quite a small ripple, and consequently there will be frequent racing, but not violent, as is the case when the ship is in a head sea-way, whether pitching or not. If there is pitching as well, there will be periods when the racing will be extremely violent, as the screw will be almost entirely freed from the water immediately following a complete submergence. Hence, as a rule, it is much better to have a small propeller which can be, and generally is, thoroughly submerged than one of larger diameter, notwithstanding the claim of the latter to superior efficiency.

CHAPTER V

PADDLE WHEELS

The paddle wheel as a marine propeller is still in use, and as it has certain qualities of its own which cause it to be valuable under certain circumstances, it is likely to continue in favour in spite of other characteristics which are against it when compared with the screw.

To-day it is employed on certain services such as—

(a) *Tug-boats.*—Where quick manœuvring is absolutely necessary it can be accomplished by disconnecting the wheels and working them independently, so that while one is moving "ahead" the other may stand still or move "astern," the turning moment being then greater than that of a twin-screw boat of equal power. Paddle wheels have also a more positive action, so that motion is imparted to the ship and retardation produced when required much quicker than with screws.

(b) In *river steamers*, when draught of water is very light, there is not the same limit to the size of propeller with the paddle as with the screw; and although this is not so marked a difference as formerly, owing to Sir John Thornycroft's and Mr Yarrow's inventions, whereby partially submerged screws can be worked with a high efficiency, it remains and is only a matter of degree. The fact is that in very shallow water the paddle wheel only can be employed with advantage.

(c) *Steamers in tropical shallow waters*, where weed grows rapidly and frequently forms a serious obstruction to navigation, are better fitted with paddle wheels, as they can be cleared when clogged much more easily and quickly than screws can be. Moreover, in practice, screw boats have a greater tendency to "suck" the bottom and be retarded by nearness to the bottom than paddle boats.

(d) *Steamers making calls at piers and wharves* which have the side paddle wheel are much more manageable and handy than twin screws, and, even in fine weather, lose less time at each stopping-place consequent on having the same qualities as a tug-boat; besides which the sponsons or platforms ahead and astern of the paddle boxes permit the placing and using of mooring bollards and posts, for the spring by which the ship is swung round, and also provide means for the quick landing and embarking of passengers and parcel cargo.

(e) **Damage to propeller** in such steamers as above enumerated is, if anything, less likely in the case of a paddle wheel than a screw, and when damaged the former can be examined and repaired, without taking the ship out of water, by an ordinary mechanic such as a country blacksmith. It is true that by providing a well over the propellers, as is now frequently done in shallow draft twin-screw steamers, the screws can be examined and even removed and replaced; it is not, however, so easily effected as in the case of a paddle wheel, especially if the water is not smooth at the time.

On the other hand, paddle wheels and the engines driving them are almost of necessity heavier, and occupy more space than obtains when screws are used, as there is a limitation to the number of revolutions of a paddle wheel far below that of the screw. The engines, etc., in the case of side-wheel steamers occupy the most valuable part of the ship, and the straining action on the hull is much more severe and has to be taken at a part less able to withstand it than is the case with screw engines. The ship occupies more space in a dock or circumscribed water, and the paddle-boxes are exposed to sea and wind to an objectionable extent. These arguments do not, however, apply to the stern-wheeler, but, on the other hand, that form of paddle-boat has not the turning or manœuvring qualities of the side-wheeler. It is claimed for the side-wheeler that she is steadier in a sea-way and does not roll to the same extent as a screw, due to the gyroscopic action of the wheels and shafts, and to the fact that the plane of the paddle floats strikes the water, in rough weather, at a considerable angle from the vertical. The ship is prevented thereby from oscillating so much as she otherwise would; as sailors say, " the wheels pick her up when she rolls."

The latter argument, however, is no sufficient set-off to the serious objections to the paddle wheel in sea-going ships, and it may be taken as practically certain that it will never be employed again in them; and further, that the tendency to-day is for the screw to have a wider application and to supplant the paddle even in the services above named, where it is employed with some advantage, perhaps largely owing to the present disposition and fashion to employ the cheap and light turbine as the prime mover, which, besides possessing those qualities, permits of the use of the screw, and actually requires it to be of quite a small diameter.

On economic grounds the screw would always be preferred to the paddle, for in prime cost and upkeep, as well as in total efficiency of propeller and machinery, it has the advantage; but of late years the paddle engine has been so much improved, and such care has been taken in every way to make it and the wheels highly efficient, that the difference is not nearly so marked as in the days when the paddle man treated the screw with contempt, and the lethargy arising from scant scientific knowledge as well as self-sufficiency, kept both him and the screw man from making the advances in construction years ago that both have achieved in later times.

The original paddle wheels had fixed floats and were known later on as the "radial," in contradistinction to the "feathering" wheel, which gradually displaced it. **The radial wheel**, however, is still used where lightness, cheapness and simplicity are prime factors in determining the choice. In small light draught river steamers, especially those plying on waters remote from engineering workshops, and where even the wayside blacksmith is not known, the radial wheel obtains, and its comparatively low efficiency has full compensations. But even the radial wheel has been modified from its original form, by making the floats in steps or by making the parts of the arms to which the floats are at an angle with the inner part attached, so that the plane of the float at entry is nearer the perpendicular than it would otherwise be, as shown in fig. 20, p. 77. Of course, floats so placed are at a disadvantage on leaving the water, as they tend to scoop the water above the normal level; but in practice this action is not appreciable in the fast-running, shallow dipping wheels of the stern-wheeler, and by no means outweighs the gain in efficiency by the better angle at entry. Of course, such a wheel when running "astern" must have a very low efficiency, but as this is only at intervals and for a short time, it really does not matter. Sometimes the floats are set so that their planes are at angles with instead of parallel to the axis, so that "entry" may be more gradual and the tail race or wake may be diverted from the side of the ship. It is very doubtful, however, if there is any substantial gain made thereby; and such improvements as are possible by such means are more often hoped for than proved, as in most other cases of propeller modifications.

In early days when paddle steamers made long runs and burnt so much fuel that the draught of water was considerably less at the end of the voyage than at the beginning, the radial wheel permitted of the floats being shifted outwards so as to get a more suitable dip when the voyage was half done. In the light draught river craft of to-day, when conveying cargo or in other ways experiencing much difference of draught, the position of the float can be altered to suit the immersion of the radial wheel, which is a distinct advantage.

The feathering wheel, the idea of which was originally suggested by Hooke and patented by Elijah Galloway, was undoubtedly an improvement on the radial, inasmuch as the floats could enter the water and emerge from it without shock and with the least amount of disturbance; and throughout the immersion they are acting solely in accelerating a stream of water in the right direction for successful propulsion. Moreover, by mechanically and automatically placing the floats and maintaining them in their true position for maximum efficiency, it was possible to have a wheel of much smaller diameter than with radial floats. Perhaps this is best appreciated by noting that the paddle steamer "Scotia," the last of the famous Cunard Atlantic paddle liners, was of 4000 I.H.P. and had a radial wheel 40 feet in diameter, whereas the modern paddle steamer "Violet," of 4070 I.H.P., has a feathering wheel of only 21·5 feet diameter; and the most powerful paddle steamer ever built, the "Empress Queen," of 11,000 I.H.P., has a wheel only 17 feet in diameter. Not only is it an advantage for the ship to have the smaller paddle boxes, but to the engineer it is a distinct gain to have small wheels, for he then must and can run at a much higher number of revolutions, and is thereby able to develop the power required with smaller and, consequently, cheaper and lighter engines.

The "Scotia" ran her trials at 15 revolutions, the "Empress Queen" at over 50, while to-day, with large and small paddle steamers, 52 to 58 and even 60 revolutions is common practice; consequently as much as 8 I.H.P. is developed per ton of machinery as against 4 I.H.P. with the "Scotia."

Paddle wheels with outer rims (see fig. 18), always used in the early days, are sometimes preferred now and are even necessary for certain services, inasmuch as they guard the floats

PADDLE WHEELS. 73

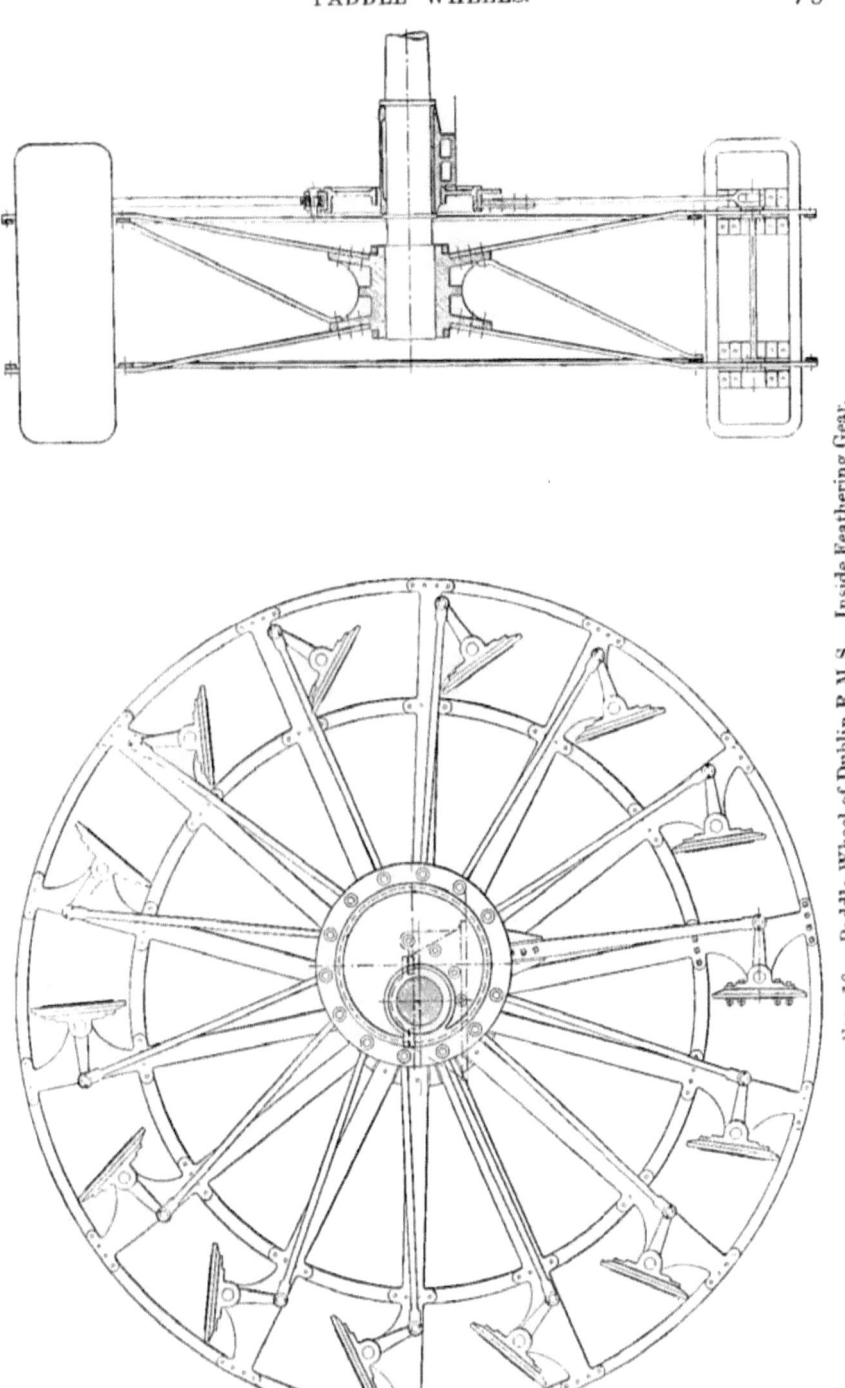

Fig. 18.—Paddle Wheel of Dublin R.M.S. Inside Feathering Gear.

from contact with floating wreckage and driftwood, whereby they would be damaged and their mechanism very likely be put out of action, if not smashed. Such a wheel, composed as it is almost exclusively of rolled bar iron subjected to the simplest of mechanical treatment, is necessarily a cheap one, and one easy to repair either at sea or in port. No one part of it is very heavy, and being *bolted together* it can be taken to pieces *in situ* when required, and every part could be made or renewed by a common smith without any special appliances. On the other hand, its extreme diameter is greater than that of the other wheel, and should floating "raffle" or wreckage get entangled in the wheel, more mischief may arise than if there were no outer rims. Fig. 18 shows the construction of such a wheel as carried out by one of the leading engineering firms forty years ago for an Irish mail steamer of high speed. The floats in it are more numerous than is the rule now, and the floats are comparatively short for their breadth.

Paddle wheels without outer rims are now almost exclusively used for high-speed steamers, on which the speed of revolution is now so high and the power of the engines so large that much greater care is necessary in both the design and the workmanship of them. Fig. 19 is the elevation of the wheel of a modern cross-channel steamer, showing how heavy are the arms required to work successfully under these modern conditions, and how carefully they are notched into the inner rim in order to get the assistance of the other arms to sustain their load when it is a maximum; with such a wheel there is more clearance for the float to oscillate and less liability of the wheel itself coming in contact with a bank when running in narrow or shoal waters, because of the absence of the outer rim. Such a wheel, however, is costly and somewhat heavy, and it cannot be so easily repaired, but the floats can be more readily removed and replaced when required.

Position of Wheels.—The "Charlotte Dundas," the earliest steamship, had a single wheel in an aperture at the stern; in fact, she may be said to have had two sterns with the wheel between them (see fig. 2, p. 9). The "Claremont" had a pair of wheels, one on each side, nearly amidships, an arrangement spoken of as "the side-wheeler" as contrasted with "the stern-wheeler" of the "Dundas." Bell's "Comet" of 1812 had two pairs of side wheels (see fig. 3, p. 11), an arrangement not then new, for ships with numerous wheels worked by other power than that of steam had been tried

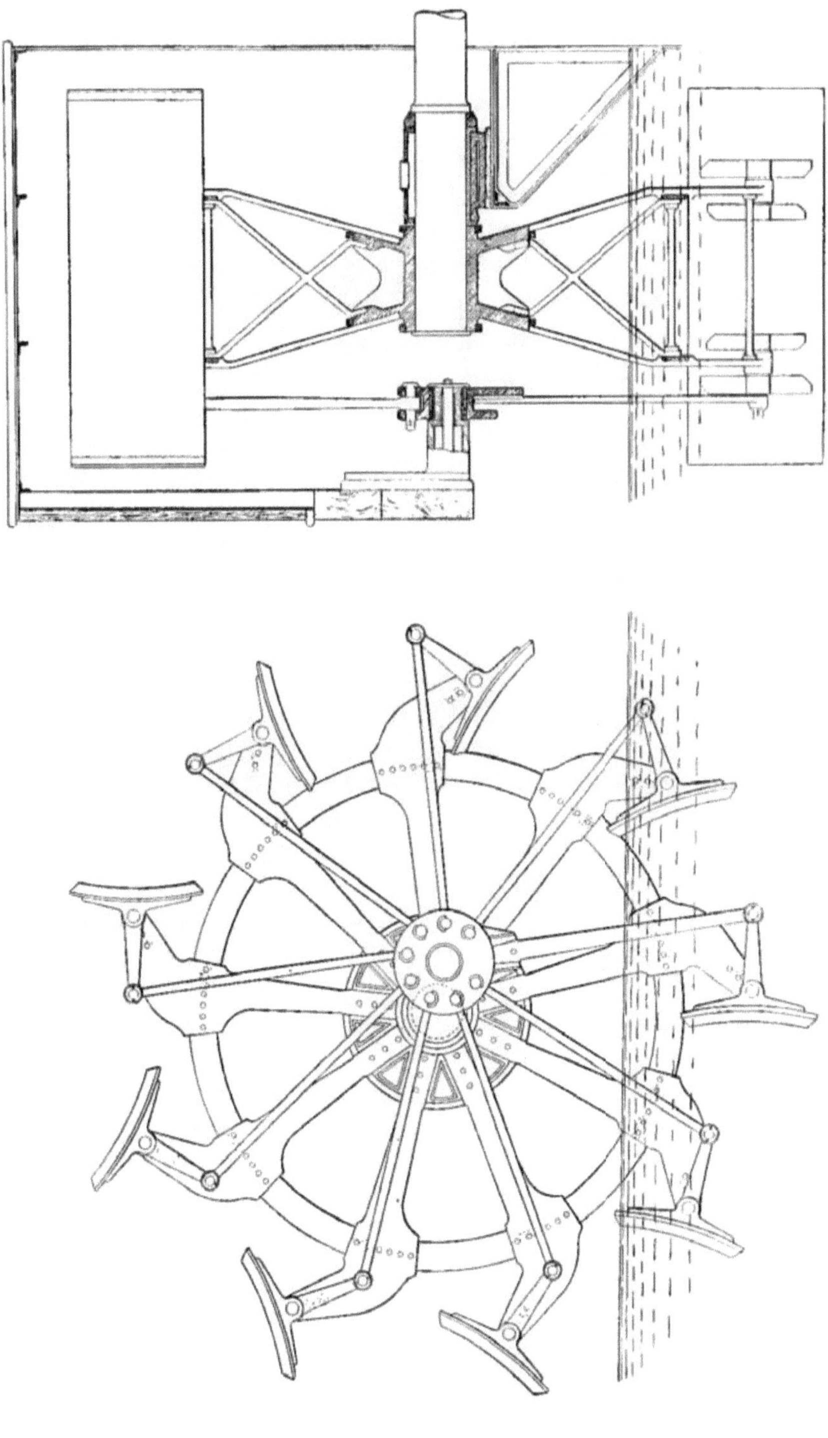

PADDLE WHEELS. 75

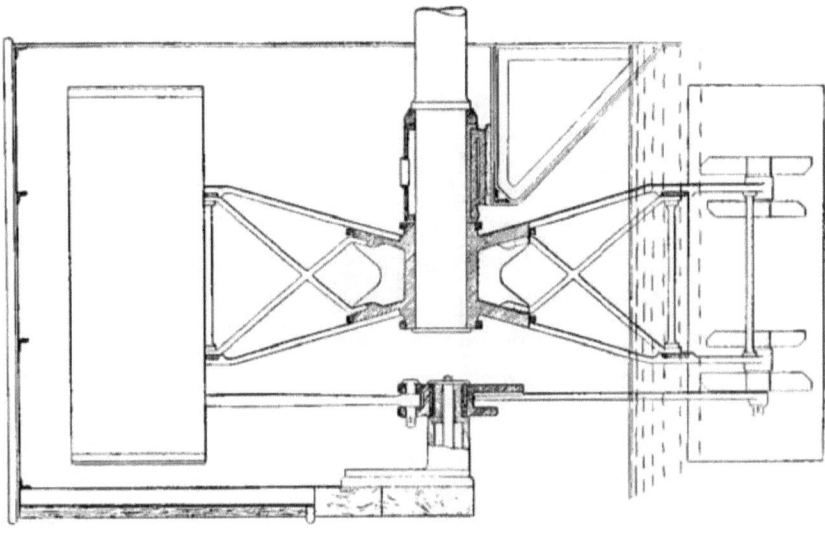

Fig. 19.—Paddle Wheel of "Normandy". (No outer rim.)

before. Almost the only other, and certainly the most striking, repetition of the two-pair wheels, was when Sir Edward Reed adopted it in the cross-channel steamer "Bessemer," 340 feet long, built in 1874. He did this to enable the centre part of the ship to be entirely devoted to the swing saloon of Sir Henry Bessemer, which was intended to prevent sea-sickness by always remaining level.

It is interesting to note that in the case of the "Bessemer" the wheels were 130 feet apart, and when going at full speed the after or following pair made 30 revolutions and the leading ones only 27 per minute, the I.H.P. exerted being about the same for both pairs of engines. It is not likely that two pairs of wheels will again be tried on a steamship.

Stern-wheelers generally have only one wheel (see fig. 20), though usually driven by an engine on each side instead of on one side only, as in the "Charlotte Dundas" (see fig. 2, p. 9). Such wheels are very often radial, and sometimes even without a box or any covering. The wheel is, however, sometimes divided into two so as to admit the engine between them. This is a convenient and economic arrangement, as may be seen in fig. 21, for the engine-room is boxed in and so confined that one man, without leaving the driving station, can keep his eye on and even attend to every part of the engine. By fitting two engines, an independent one to each, the ship is much more manageable and less liable to be at a standstill from damage to wheels, and the extra cost of such an arrangement is trifling compared with the advantages derived from it.

Three paddle wheels are sometimes fitted to the stern of large beam shallow draught ships so as to get greater propulsive power. The wheels are, of course, in line axially, and are generally so arranged that there is a crank between a pair of wheels—that is, there are two cranks between the three wheels.

A remarkable boat of this kind was built at Wiborg in Finland in 1885; she was 110 feet long, 25 feet beam and 4 feet deep, and drew only 22 inches of water. The cylinders were 14 inches and 28 inches diameter and 5 feet stroke. The paddle wheels were 12 feet diameter, and when running at 37 revolutions per minute attained a speed of 11 knots with 350 I.H.P.

Single wheel amidships is generally associated with the name of Captain Dicey, who advocated the system so successfully that a cross-channel steamer named "Castalia" was built and tried on the Dover-Calais station in 1874. On her proving a failure his backers got a larger and more powerful one on the station, named

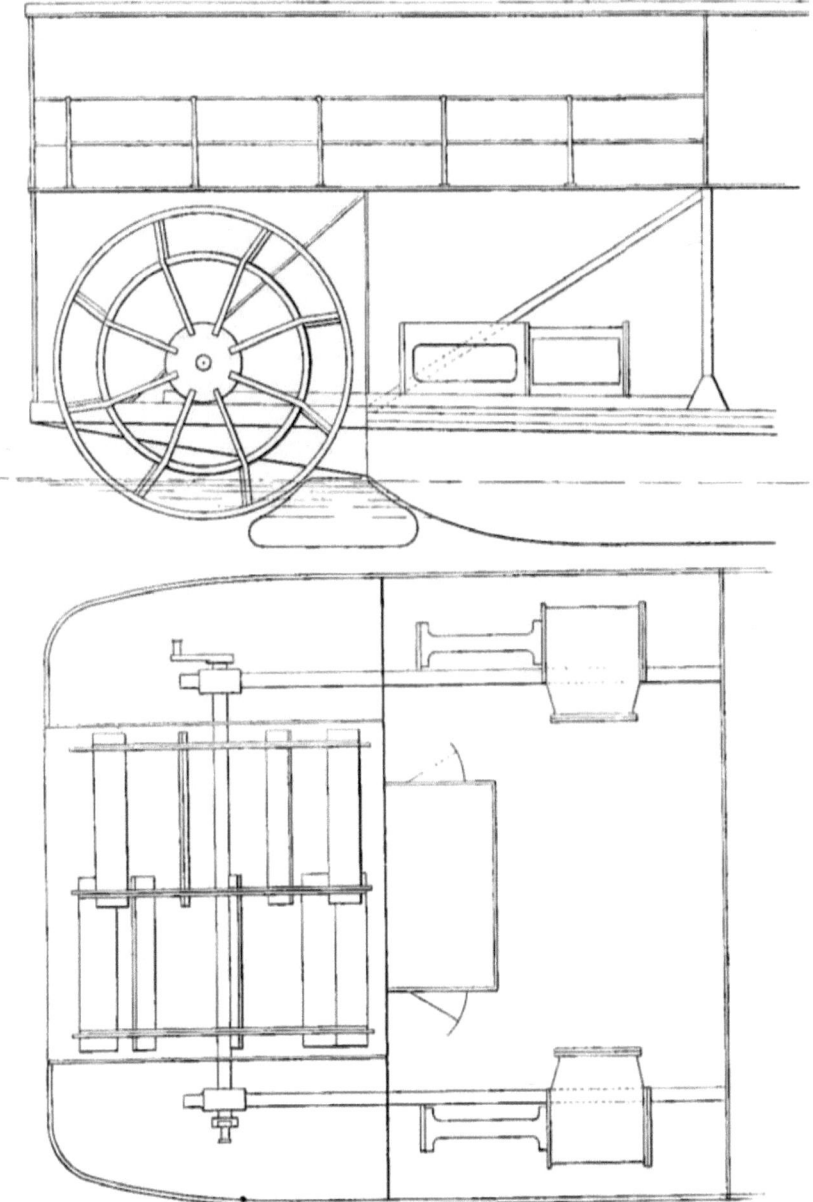

Fig. 20.—Stern-wheel Steamer. Single Wheel.

the "Calais Dover," but she too proved a failure. The fact is, each ship really consisted of two hulls set far enough apart to permit of the working of the one huge paddle wheel and connected above water so as to appear as a single ship. The immersed skin was about double that of an ordinary ship, and the channel between the hulls,

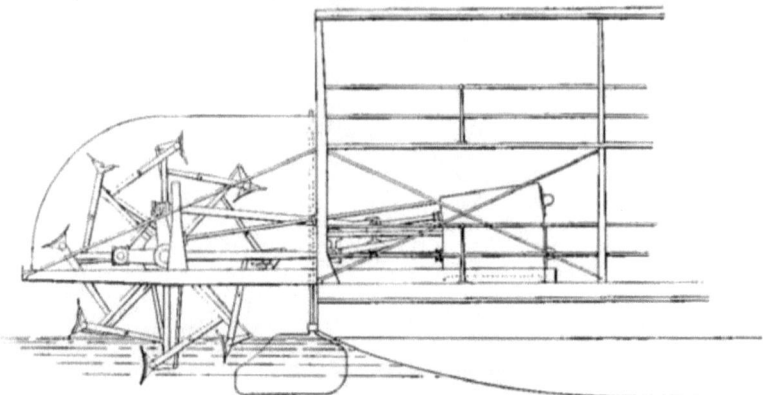

Stern Wheeler.   Feathering Floats.

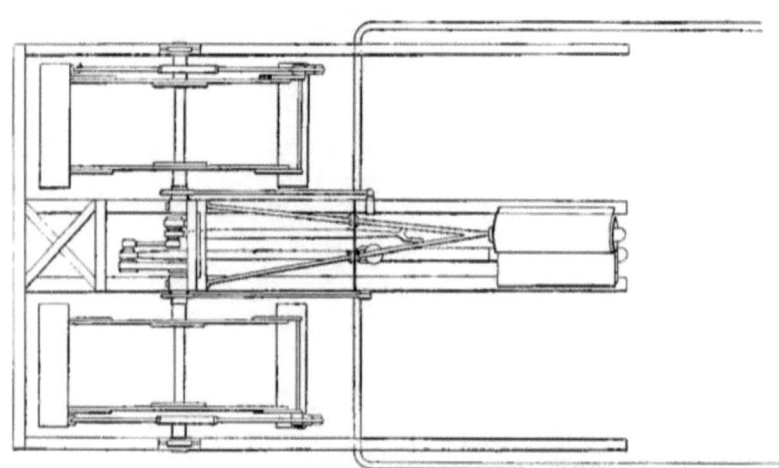

Fig. 21.—Stern Wheel of the "Endeavour."   Pair of Wheels.

subject to the inflow to and the wake from the wheel, offered great resistance, as did also the cover of this channel in bad weather.

Central-wheelers were tried in quite early days, and on account of the protection afforded to the wheel and the extreme beam tending to steadiness in a sea-way, it was no unnatural thing that the arrangement should attract attention for cross-channel service in the days when it was thought the screw was quite unsuitable for such service.

PADDLE WHEELS. 79

TABLE VI.—EXAMPLES OF PADDLE WHEEL STEAMERS HAVING SIDE WHEELS WITH FEATHERING FLOATS.

| Name of Ship. | Dimensions, Feet. | | | Displacement. Tons. | Results of Trials. | | | | $\frac{D_3^2 \times S^3}{I.H.P.}$ | Paddle Wheels. | | Pr.H.P. | Tow Rope H.P. | $\frac{Pr.H.P.}{I.H.P.}$ | $\frac{Tr.H.P.}{I.H.P.}$ |
	Long.	Beam.	Draft.		Speed.	Revs.	Slip per cent.	I.H.P.		Diameter.	Floats.				
										Feet.	Feet.				
"Connaught"	338	35·0	13·0	1900	17·80	28·0	24·2	4751	182·0	27·0	11·3 × 4·0	3730	2525	0·784	0·532
"Violet"	300	33·0	10·7	1485	19·50	35·0	15·0	4070	237·0	21·7	11·0 × 4·5	2566	2500	0·616	0·614
"Princess Marie"	275	34·8	11·0	1815	17·12	32·75	19·7	3543	...	21·0	12·0 × 4·17	2654	1772	0·749	0·500
"Ireland"	366	38·0	11·0	2160	23·00	27·0	14·2	6300	204	23·0	12·7 × 5·75	4116	4056	0·653	0·644
"Invicta"	312	33·6	...	...	17·7	33·8	21·5	3907	...	21·6	9·75 × 4·0	2627	...	0·658	...
"G.S."	260	28·0	5·8	663	18·25	47·0	26·5	2680	172·4	17·0	9·75 × 3·5	...	1307	...	0·488
"Normandy"	230	27·6	6·0	640	18·26	41·0	15·5	2520	178·5	16·3	9·5 × 3·6	1236	1147	0·500	0·455
"Egypt"	202	21·0	6·25	435	14·78	42·0	15·8	882	207·0	13·5	8·0 × 2·5	506	479	0·575	0·544
"Channel"	199	21·5	6·0	450	14·80	38·0	17·3	1108	185·0	15·5	8·0 × 3·0	738	728	0·666	0·652
"Essex"	175	23·0	5·5	370	14·0	5·50	26·6	780	180·1	11·2	7·0 × 2·8	361	341	0·463	0·437
"John Penn"	172	18·7	6·8	250	15·3	4·00	16·7	890	193·0	14·8	7·2 × 2·9	661	411	0·844	0·501
"L.C.C."	130	18·5	2·8	125	11·5	63·0	32·6	374	112·0	8·75	6·1 × 1·7	341	107	0·910	0·316
"Diligence" (Tug)	144	27·2	10·7	695	11·75	35·8	30·2	1376	92·3	15·12	9·0 × 3·3	956	311	0·695	0·227
"Osborne," H.M.S.	250	36·2	14·8	1850	16·6	27·0	17·8	3209	215·1	24·16	11·5 × 3·6	1754	1628	0·547	0·509

CHAPTER VI.

DIMENSIONS OF PADDLE WHEELS.

The diameter of a paddle wheel depends on the speed of the ship, the revolutions of the wheel, and the rate of slip. Let D be the effective diameter of the wheel, which, for purposes of calculation, shall be taken to the float "centres" of a feathering wheel and the middle of opposite floats of a radial wheel;

A the area of one float in square feet;
V the velocity of the wheel at float centres, etc., in feet per second;
S the speed in knots;
v the velocity of the ship in feet per second;
s the speed of the ship in knots;
e the efficiency of the machinery;
E the efficiency of the machinery and propellers;
R the revolutions per minute of the wheels;
T_R the tow rope resistance of the ship in lbs.;

(1) The power delivered to the wheel $N.H.P. = I.H.P \times e$;
(2) The power delivered by the wheel $P_R.H.P. = I.H.P. \times E$;
(3) The stream of water from wheel per second $= A \times V$ cubic feet;
(4) The weight of water projected $= A \times V \times 64$ pounds;
(5) The mass of this water $= A \times v \times \dfrac{64}{32} = 2A \times V$;

(6) The acceleration given to the water $= V - v$ per second; then the pressure on the float or the thrust on shaft bearing from reaction at float $= 2A \times V(V - v)$.

(a) The work done = thrust multiplied by speed of ship in feet per minute
$$= 2A \times V(V - v) \times 60v = 120A \times V(V - v)v.$$

(b) $P_R.H.P. = \dfrac{120A \times V(V - v)v}{33,000} = I.H.P. \times E = \dfrac{A \times V(V - v)v}{275}$.

Now since $V = \dfrac{\pi D \times R}{60}$ and the speed of the ship is a known

DIMENSIONS OF PADDLE WHEELS.

quantity, it is easy to determine from the fundamental equation (a) the relation between D and A.

$$D = \frac{60V}{\pi R} = \frac{v+(V-v)60}{\pi R}.$$

$$D = \frac{19}{R}\{v+(V-v)\} \text{ or } \frac{19}{R}V.$$

That is, the diameter of wheel = speed of ship *plus* slip (both in feet per second) $\div \pi \times$ the revolutions per minute.

Examples.—To find the diameter of the wheel for a ship whose speed is to be 20 knots when running 50 revolutions per minute with a slip of 20 per cent.

Here $v = \dfrac{20 \times 6080}{60} = 2027$ feet per minute, or 33·8 feet per second.

Slip $V - v = \dfrac{20V}{100}$; or $\dfrac{V80}{100} = 2027$; then $V = 2534$ feet per minute, or 42·2 feet per second.

$D = 2534 \div 3\cdot1416 \times 50 = 16\cdot14$ feet.

Reaction on the float of a paddle wheel whose diameter is D at R revolutions per minute and the slip 20 per cent. The float area being A, the speed of wheel per minute is $\pi D \times R$ or $\dfrac{\pi DR}{60}$ per second;

the acceleration is $\dfrac{20}{100} \times \dfrac{\pi DR}{60} = \dfrac{\pi DR}{300}$;

weight of water moved per second $= \dfrac{A \times \pi DR}{60} \times 64$;

the mass of water moved per second $= \dfrac{A \times \pi DR}{60} \times \dfrac{64}{32} = \dfrac{A \pi DR}{30}$;

thrust in pounds $= \dfrac{A \times \pi DR}{30} \times \dfrac{\pi DR}{300} = \dfrac{A(\pi DR)^2}{9000} = A\left(\dfrac{DR}{30}\right)^2$;

the speed of the ship is $\dfrac{80}{100} \times \pi DR = 2\cdot5$ DR feet per minute;

work done $= A \times \left(\dfrac{DR}{30}\right)^2 \times 2\cdot5 \text{ DR} = \dfrac{A \times (DR)^3}{360}$;

E.H.P. $= \dfrac{A \times (DR)^3}{360 \times 33,000} = \dfrac{A \times (DR)^3}{11,880,000} = A \times \left(\dfrac{DR}{228}\right)^3$;

that is, the area of one float = E.H.P. at one wheel $\times \left(\dfrac{228}{DR}\right)^3$

$$= \text{I.H.P.} \times \text{E} \times \left(\dfrac{228}{DR}\right)^3.$$

Example.—To find the area of a float of a wheel whose diameter is 22 feet, the revolutions 36 per minute, the I.H.P. per wheel 2100, and the efficiency 0·66.

$$\text{Area} = 2100 \times 0.66 \times \left(\frac{228}{22 \times 36}\right)^3 = 33.5 \text{ square feet.}$$

Area of Floats.—In practice the following rule holds good. Let I.H.P. be the total power developed by the engine, then each float should have the area A in square feet, D being the diameter in feet to float centres and R the revolutions.

Rule.—$A = \text{I.H.P.} \times E \times \left(\frac{228}{D \times R}\right)^3 \times F.$

For high-class machinery and feathering floats . $E = 0.7$
For ordinary   ,,   ,,   ,, . $E = 0.65$
For high-class machinery and radial floats . $E = 0.60$
For ordinary   ,,   ,,   ,, . $E = 0.55$
For stern-wheelers with one wheel only. Radial . $F = 0.70$
For   ,,   ,,   ,,   Feathering $F = 1.15$
For side-wheelers with two wheels. Radial . $F = 0.40$
For   ,,   ,,   ,,   Feathering . $F = 0.60$

The pitch of the floats—that is, their distance apart—must depend largely on the speed of the wheel. Each float sweeps the water before it as shown in fig. 14, leaving a hollow behind into which the oncoming water flows by gravity; consequently there is a slope of water always behind each float, while the surface at which the next float will touch will be below that of still water; that is, below the plane of the flotation of the ship when at rest.

It is easy to calculate how far the back of a float may be bare from the common formula $v = 2gh$; taking g at 32 and v as the velocity imparted by the float in feet per second, we have $h = \frac{v}{2g}$.

Example.—To find the amount of the baring of a float moving at a velocity 50 feet per second when the slip is 20 per cent., the acceleration is 20 per cent. of 50 or 10 feet. Then the head

$$h = \frac{10^2}{2 \times 32} = \frac{100}{64} = 1.56 \text{ feet.}$$

That is, the float in question, if its dip is such as to bring its inner edge on the water level, will have its back bare for a depth of 1·56 feet when the ship is moving at a speed of 23·6 knots.

DIMENSIONS OF PADDLE WHEELS.

Number of Floats.—*An old rule for number of floats is*

$$\text{Number of floats} = 60 \div \sqrt{R}$$

So that with constant revolutions the larger wheel has the greatest pitch of floats.

Proportions of floats of a feathering wheel differ from those of a radial, thus:—

In a radial wheel ratio of float length ÷ breadth = 4 to 5.
In a feathering wheel „   „   „   = 2·8 to 3.

Originally, floats were made usually of English elm. To-day that timber, as well as Canadian elm, makes a satisfactory board; in fact, what is wanted is a strong, tough wood which stands the action of water and which does not easily splinter or warp; anything having these qualities and not too heavy will do for the purpose. For inshore working tug-boats even pitch pine has been used; but this splits too easily, and is, moreover, a fairly heavy wood.

Thickness of wooden floats on a radial wheel was usually about one-eighth of their breadth, those of a feathering wheel one-twelfth their breadth. Ratio of breadth of float to thickness is for radial 8, feathering 12.

The large floats are made in more than one piece, and in any case it is a good thing to through-bolt them, although it is urged that in doing so they were weakened against cross bending by having the holes. A better method of tying them together is therefore by bolting through and through flat bars, or even angle bars, fitted across them, as the float is also stiffened thereby.

The edges, especially the outer ones, of wooden floats are always carefully bevelled, as are also the ends, both being on the back sides.

Steel floats are often used to-day instead of wood, because they offer less resistance at entry and exit as well, and are easily kept with a smooth surface by the application of black varnish or enamel paint. Such floats may also be made curved as shown in fig. 19, and are stiffened by angle bars as well as by the feathering gear. Steel plate floats are heavier than wood, and in the opinion of the old engineers, whose experience was largely with paddle steamers, the latter is preferable, inasmuch as a wooden float is incapable of getting bent out of shape as a steel one is, and therefore cannot do mischief to the boxes, to the wheel, the gear itself, and other parts. They are more easily removed than a bent steel one, and the ship's carpenter can always make a new one as well as repair an old.

Construction of the paddle wheel is seen and can be studied by referring to figs. 18 and 19. It consists of a large boss with conical flanges at each end to suit the splay out of the arms. The old radial wheels had a boss with parallel flanges and the larger ones with a flange in the middle to suit the three sets of arms, rims, etc.; and as there was no feathering gear to arrange for, the boss was as long as the wheel was broad at the rims.

Paddle-wheel bosses were made of cast iron of the best and toughest quality, and great care was taken to avoid splitting at the flanges. To guard against the contingencies of hidden cracks, and to fortify the boss to withstand the heavy shocks it gets in a sea-way, the cast-iron boss had been shrunk on the edge of each flange, and on spigot ends formed around the shaft-hole rings or tyres of wrought iron, and latterly of steel. (See fig. 19.)

Since steel castings have been available for the purpose, it has been customary to make these bosses of that metal, taking care that they were carefully annealed and "let down," so as to have no initial stresses due to cooling in the broad flanges.

Formed on the flange are fillets, between which the arms are carefully bedded so as to take all shear off the bolts. In the same way these fillets are fitted to the facings for the diagonal ties, so that there may be as little shear as possible on their bolts. The boss is carefully bored out so as to fit on the shaft end; and although it was the old practice to bore the hole parallel, which was an improvement on the system preceding it, wherein it was rough and the boss was "staked" on with four stakes or long wedges, it is better now to make the shaft end slightly taper, say one in 48, and key it on tightly with two good-fitting keys at right angles.

Let d be the diameter of the inner journal of a paddle-wheel shaft as found from the formula $d = \sqrt{\dfrac{\text{I.H.P.}}{\text{R}} \times \text{F}}$.

(a) For an engine with a single cylinder . . . $F = 80$
(b) For an engine with two cylinders and cranks coupled
    at right angles . . . . $F = 58$
(c) For an engine with two cylinders and intermediate
    shaft . . . . $F = 50$
(d) For an engine with two cylinders, solid crank shafts
    at right angles . . . . $F = 55$

DIMENSIONS OF PADDLE WHEELS.

Thickness of cast-iron boss around shaft	$= 0{\cdot}28 \times d$
,, ,, cast-steel ,, ,, ,,	$= 0{\cdot}20 \times d$
Diameter of outer flanges	$= 4 \times d$
Thickness of cast-iron flanges	$= 0{\cdot}15 \times d$
,, ,, cast-steel ,,	$= 0{\cdot}11 \times d$
Breadth of two hard steel keys, each	$= 0{\cdot}18 \times d + 0{\cdot}25$ in.
Thickness ,, ,, ,,	$= 0{\cdot}09 \times d + 0{\cdot}25$ in.

N.B.—All shafts above 8 inches diameter should have two keys of about the size given above. When the diameter is less than 8 inches, one key is sufficient, and it should be not less than given by the above.

Paddle-wheel arms in small radial wheels are of flat bar wrought iron or mild steel (see fig. 20). In larger wheels ordinary bar can only be used if the power is comparatively small. In case of large power and of all feathering wheels, the arms are forgings, formerly of iron, but now always of steel, as being cheaper and easier to obtain (see fig. 19); the whole arm can be smithed from one piece instead of by welding on the brackets for floats.

Let N be the number of floats on a wheel and each float be supported by a pair of arms whose breadth is b and thickness t. Then, if the framework of the wheel is properly designed and fitted so as to distribute the load, the resistance to bending of each arm $= f(t \times b)^2$ and—

Total resistance $= 2nf(t \times b^2)$.

The shaft inner journal has to work against this same resistance, so that, if it be taken as P,

Then $P \times \dfrac{D}{2}$ is the torsion on the shaft, and also equal to the bending moment on the arms at the centre.

The torque on the shaft $= \dfrac{\pi d^3}{16} \times f$;

then $\qquad P \times \dfrac{D}{2} = \dfrac{\pi d^3 \times f_1}{16} = 2nf(t \times b^2)$;

that is, $\qquad t \times b^2 = \dfrac{f_1}{f} \cdot \dfrac{\pi d^3}{2 \times 16} = {\cdot}0982 \times d^3 \times \dfrac{f_1}{f}$.

Ratio of b to t at boss, 5, and near rim, 3·5; when there is only an inner rim the ratio of b to t outside the rim will be 6 to 7, and for mild steel $\dfrac{f_1}{f} = 0{\cdot}8$.

It follows, then, that $b^2 \times \dfrac{b}{5} = 0.0982 d^3 \times 0.8$ or $b^3 = 0.393 d^3$

$$b = d \sqrt[3]{0.393}.$$ Therefore :—

breadth of arms near boss $= 0.732 \times d$;
thickness ,, ,, $= 0.146 \times d$; and
breadth of arms near rims $= 0.511 \times d$.

It must always be borne in mind that the stresses on a wheel in a tug or other open sea-going ship are far greater in proportion to the driving power than is the case in ships running always in smooth water, for there is, in addition to the shocks from the waves, the liability of the whole available power to be transmitted to the one "lee" wheel, the other, or "weather" one, being quite out of water. Unless the floats have been designed for such an emergency the engine will race and then probably develop the full power; for otherwise, in such rough weather the engines would be probably "eased down." If d is determined to suit such conditions the I.H.P. in the formula, p. 84, instead of being one-half, will be three-quarters the whole power developed. Consequently, by taking d as a basis for calculation, the wheel will resist to the same extent as the shaft. It may also be noted that for smooth-water steamers a small factor of safety is sufficient. In these cases $\dfrac{f_1}{f}$ may be taken at 0.65 to 0.60 instead of 0.8.

Paddle-wheel rims are of flat bar iron or mild steel in section, depending on the pitch of the arms and the size of the parts of the arms near them. In the old wheels, where the floats were closer together than obtains now and the number of floats was consequently large, the rims were generally of about the same section as that of the arms; in fact, small radial and feathering wheels were often made of one section throughout. In the case of a modern feathering wheel with two rims the following holds good :—

Breadth of outer rim	$= 0.40 \times d$
,, inner ,,	$= 0.40 \times d$
Thickness of outer rim	$= 0.08 \times d$
,, inner ,,	$= 0.10 \times d$
Inner ring, when no outer, breadth	$= 0.50 \times d$
,, ,, thickness	$= 0.14 \times d$
Diameter of bolts throughout	$= 0.12 \times d$, double rims.
,, ,, ,,	$= 0.15 \times d$, single rims.

Tie bars are placed diagonally from the inner rim between each pair of arms to the flange of the boss, and cross-bars to arms behind the floats.

Diameter of diagonal ties $= 0\cdot18 \times d + \frac{1}{2}$ inch.
" " cross-bars $= 0\cdot18 \times d + \frac{3}{4}$ inch (made hollow)

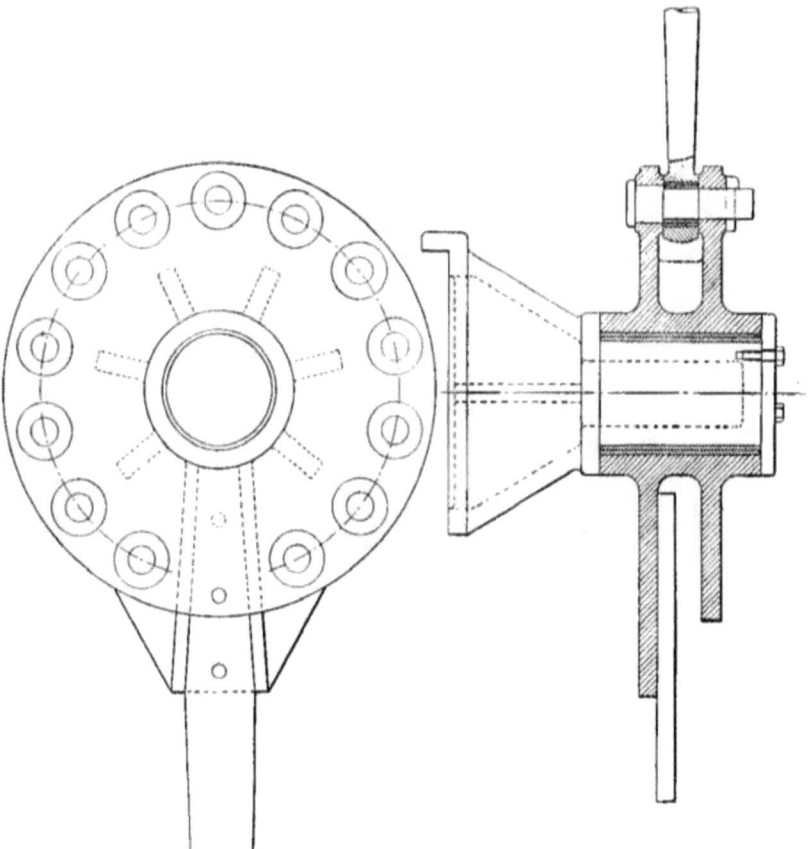

FIG. 22.—Details of Feathering Gear of a Paddle Wheel.

It need hardly be said that after making these calculations for a proposed wheel the nearest standard section sizes must be chosen and dimensioned accordingly on the drawings.

Feathering gear as now fitted to wheels differs very little from that found on the Morgan wheel as designed by Elijah Galloway seventy years ago. There have been large numbers of other methods of effecting the feathering, some of which were as ingenious, some as simple, and some even as effective; but none of them have survived, because none possessed all these qualities in the same degree as Galloway's invention, which is shown in fig. 19, the feathering of the floats being effected in that case by their levers being connected to and revolving about a pin fixed to the sponson beam of the paddle box set eccentrically to the paddle-shaft axis. (See fig. 22.)

This method is a very convenient one, as it takes the gear to a place where it forms no obstruction. Unfortunately, however, the eccentric pin is affixed to something which at all times is liable to spring and never forms the rigid base desirable for it, for it sometimes

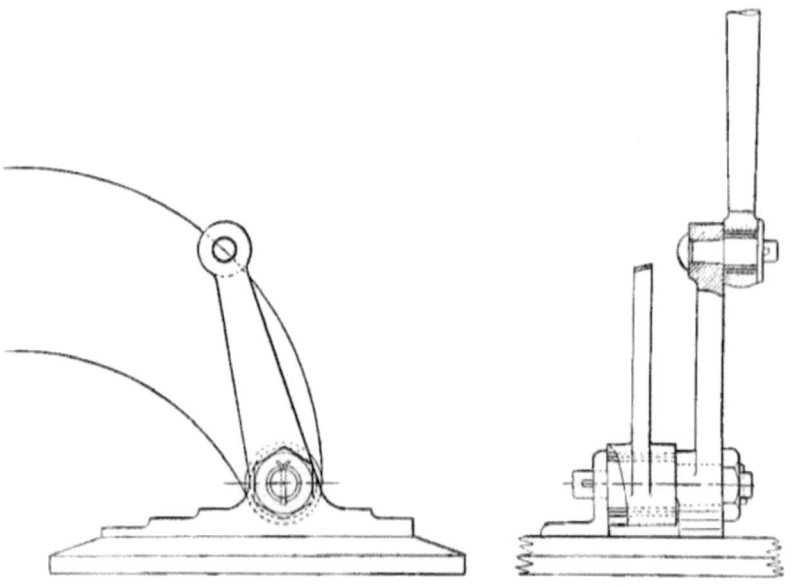

FIG. 23.—Details of Float Bearing, etc., of a Feathering Wheel.

gets so badly bent or sprung by contact with a quay wall or dolphin as to damage the feathering gear and put the wheel out of action. It was therefore arranged many years ago, and by some engineers adopted as their general practice, to fit an eccentric to the outer end of the shaft main bearing of such a diameter as to permit of feathering sufficiently and clearing the bearing and its base, as shown in fig. 18. In this case the frictional resistance of the large hoop is necessarily greater than that of the pin carrier of the older method. It can, however, be lubricated with oil and attended to in a way that was not possible with the pin on the sponson beam.

The working parts of the feathering gear of a paddle wheel should

DIMENSIONS OF PADDLE WHEELS.

have the pins cased with hard bronze and the holes in which they fit bushed with lignum vitæ. (See fig. 23.) The feathering pin should be treated in the same way, as it is always being drenched with water, so that no effective oiling is possible.

White metal bushes (Fenton's) with hard steel pins can be employed satisfactorily, especially in sandy or muddy water.

Dimensions of Float Fittings.—The diameter of float gudgeons = $0.1 \times$ breadth of floats $+ \frac{1}{2}$ inch.

Length of bush for gudgeon = $1.40 \times$ diameter of gudgeons.

The diameter of radius rod pins = $0.50 \times$ diameter of gudgeons.

The king rod should have a double jaw support to its pin and the float gudgeons strengthened to take the load of driving the feathering gear.

The locus or path of the paddle float "centre" is shown in fig. 24. The position of the float or the angle it makes with the vertical at each point is clearly seen throughout the whole course, but the most im-

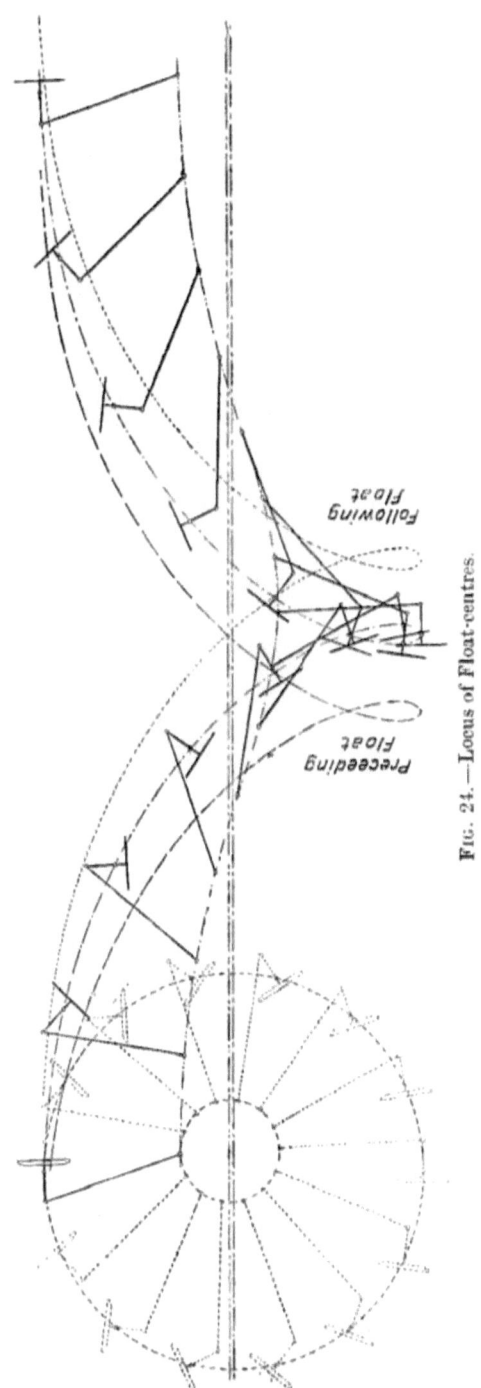

Fig. 24.—Locus of Float-centres.

portant and interesting period is that during which the float enters, passes through, and emerges from the water. It is evident, on examining the diagram, that the aim of some of the older engineers to have a float vertical throughout that stage was futile, and such an arrangement would have detracted very much from the efficiency of the propeller. To enter and leave the water without shock is necessary to efficiency, and this could only be accomplished by causing the float to be as nearly tangential to the *locus* as possible at those points.

The path is easily constructed by remembering that while the wheel is turning round on its centre it is advancing with the ship; and since the tangential velocity is at a higher rate than the horizontal advance of the ship, there will be the loop at the immersion period, due to the differences in direction and velocity, so that the horizontal difference in position of a point on the path is the measure of the acceleration imparted to the water in the time taken in effecting the movement.

CHAPTER VII.

HYDRAULIC PROPULSION: INTERNAL PROPELLERS AND JET PROPELLERS.

As early as 1698 Savery patented his engine, whereby large quantities of water could be drawn into it and expelled at a comparatively low velocity with a moderate consumption of steam by means of the simplest of appliances. It is true that it was necessary to have a boiler capable of producing steam at a pressure in excess of the atmosphere, but this he must have had under any circumstances. His engine, in fact, was the original and rudimentary form of the modern pulsometer, but it was not automatic.

Not till 1729, however, was it recognised as a fundamental principle that the propulsion of a ship was effected by projecting a stream of water in the opposite direction to her motion, for in that year John Allen took out a patent (see p. 4) for propelling a ship with a machine like Savery's, but, strange to say, instead of steam he proposed to use what we now term "internal combustion" gases to obtain the pressure for ejecting the water.

Had this principle of propulsion been known to and recognised by Savery there might have been a *steam*ship, something like that shown in fig. 1, in A.D. 1700; and the self-propelled ship might have come into general use a hundred years earlier than it actually did.

The simplicity of this means of propulsion will always incite a strong attraction for it, and especially will it be so to those men who constitutionally hate intricacy and complications, and who consequently condemn any deviation from it in design and construction, however good such may be both in principle and practice.

A steamship propelled by such means was actually built and tried so late as 1876 by Dr Fleisher in Germany. She was named the "Hydromotor," and was of 105 tons, the length being 110 feet, the beam 17 feet, and the draught of water 6·2 feet. It is said that she attained a speed of nine knots with the equivalent of 100 I.H.P. The two receivers in this case were cylindrical, lined with wood, and in each was a wooden piston floating on the surface; they worked alternately as a pulsometer does, and ejected about 12 cubic feet of water per second at a velocity of 66 feet. Such success as was attained with this ship was apparently insufficient to encourage the building of others. It should be noted, however, in passing that Joseph C. Napier patented a very similar arrangement in 1817.

Hydraulic propulsion was a very proper designation for this method of propelling ships, inasmuch as there was absolutely no mechanical instrument which might be worked on any form of propeller, however much it might differ from the paddle or screw. When, later on, ships were built and their propulsion attempted by such means, the stream issuing was small in section, and, with high velocity, it had another descriptive name, viz., **jet propulsion**, which was so called because the stream issued in jets of high velocity from either side of the ship at or near the stern, instead of through one large orifice at a low rate of flow.

Allen's system of 1729, already referred to, was succeeded by that of his new patent of 1830, wherein he claimed to use a Newcomen engine to pump the water through the nozzles, he having no doubt found that his internal combustion machine, like so many new, immature, and untried inventions, could not be relied on to successfully compete with the older and improved steam engine, either in economy or continuous good working. But even with the help of the Newcomen engine there must have been still other impediments, for little or nothing more is to be found in the records of Allen or his inventions. Eight years later, however, we find that Bernoulli proposed a system of jet propulsion very like Allen's, so that it may be that he had learned from London what Allen had done, and perceived in his experiment the germ of greater things which were to be achieved when the defects of Allen's gear and apparatus were removed, or perhaps he hoped that when modified under the brighter guiding light of Continental science they would be rendered less harmful and approach nearer to efficiency.

James Ramsay's system of 1792 differed from Allen's by his proposing to use a *centrifugal pump* instead of a common reciprocating pump. He either could not get such a pump made, or thought so little of his invention as to abandon it for an ordinary vertical pump, for in the following year, 1793, he had built and tried on the Thames a boat of considerable size driven by a common pump 24 inches in diameter. The jet or orifice was, however, only 6 inches in diameter, so it is not very surprising that the speed of the boat was only 4 knots. This, however, was about as good as Symington's paddle boat of 1788. Had he tried a centrifugal pump and a 10-inch orifice the early history of the steamship might have been different.

Internal propellers seems to be a proper and distinguishing name for the pump pistons and the centrifugal pump impeller of Ramsay. It certainly is appropriate to that of William Hales, who in 1827 to 1836 employed a vertical helix or screw turning in an enclosing cylinder as the pump to draw in water from the bottom and expel it through the stern.

John Ruthven's system, as patented in 1849, was, however, the best known and most carefully thought-out and extensively tested of the various systems for obtaining motion without the use of an exposed screw or of the still more exposed paddle.

Just as F. P. Smith was by no means the originator of the screw propeller, nor really the first who had taken out a patent for a propeller of the kind, yet our veneration is due to him for the perseverance with which he acted and proved the merit of his invention, and our thanks as well as his to Mr Wright for providing the means for doing it; so Ruthven, while entering the field so late with his plan for hydraulic or jet propulsion, was the first to succeed in having the system thoroughly tried on a large scale; it was, however, seventeen years after the patent was taken out that the Admiralty was persuaded to build a special vessel of 1160 tons displacement, to satisfy the public whether Ruthven's claims were or were not so well founded as his backers maintained.

Referring to the extract from Ruthven's patent No. 12,739, set out on p. 24, it will be seen that he claimed to use a centrifugal wheel with curved blades—in other words, a particular form of the centrifugal pump which Ramsay had claimed in 1792.

Fig. 25, p. 94, shows a rough plan of Ruthven's propeller and pipes as supplied to H.M.S. "Waterwitch" in 1866. The terminal arms or elbows shown to be fitted on each side could be swivelled round so as to deliver the stream of water right aft, right forward, or at any intermediate position; and as the gear for moving them round was operated from the bridge, she could be steered as well as have her speed regulated by the officer on the bridge without reference to the officer in the engine-room, much to the delight of the former, as all the latter had to do then was to keep the engines running at full speed. The "Waterwitch" was put into competition with her sister ship H.M.S. "Viper," as may be seen by referring to the figures in Table VII., wherein are to be found recorded results of the best trial of each ship. An examination of all the trials, however, shows that there was really not much to choose between the two ships, and that the efficiency of both was wretchedly poor, as may be concluded by comparing them with the performances of two single-screw ships of

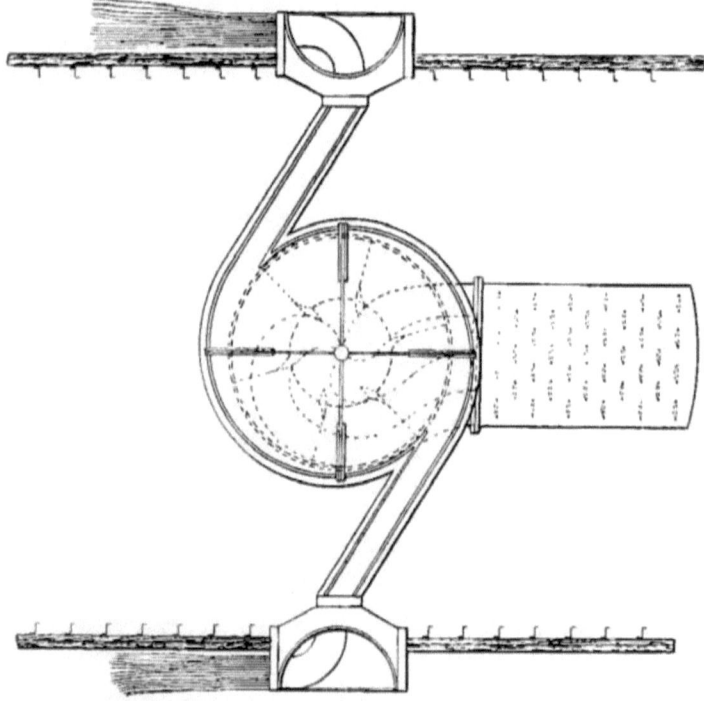

Fig. 25.—Ruthven's Hydraulic Propeller, as in H.M.S. "Waterwitch."

similar size and shape made since, and with one single-screw naval ship somewhat older; all three were of about the same speed. Two other ships, one a twin-screw and one a paddle steamer of higher speeds, are added, to show that even then their efficiency is much greater than that of the "Waterwitch," in spite of the increased and somewhat high speed for their length.

The "Waterwitch" speed was low even for 1866; to-day no kind of naval ship would be of any use whatever at so slow a speed. To quicken her machinery up sufficiently for the speeds of to-day would be an impossibility; and if even 12 knots only had been required from the "Waterwitch" it could not have been obtained, except by increasing her draught of water to permit of the heavier boilers, etc., and the enlarged water-passages and chamber for the wheel.

TABLE VII.—TRIAL OF THE HYDRAULIC PROPELLER H.M.S. "WATERWITCH" COMPARED WITH THOSE OF OTHER SHIPS OF ABOUT THE SAME SIZE.

Name of Ship.	Displacement. Tons.	Dimensions, Feet.			Trial Results.			
		Length.	Beam.	Draught of Water.	Speed.	I.H.P.	$\frac{D^{\frac{2}{3}} \times S^3}{I.H.P.}$	
H.M.S. "Waterwitch"	1161	162·0	32·0	11·17	Knots. 9·30	760	117	Hydraulic propeller, Ruthven's.
H.M.S. "Viper"	1180	162·0	32·0	11·83	9·58	696	141	Twin-screw propeller; expansive engines.
H.M.S. "Chameleon"	1136	185·0	33·2	13·50	10·21	584	196	Single-screw propeller; horizontal engines.
S.S. "Bolama"	1286	200·0	30·0	12·1	10·80	661	228	Single-screw propeller; vertical triples.
S.S. "Polo"	1035	160·0	25·0	13·0	9·50	430	203	Single-screw propeller; quadruple triples.
T.S. "Ivy"	1080	204·0	34·0	10·3	13·20	1018	240	Twin-screw propeller; vertical triples.
P.S. "M...N"	1378	245·0	32·0	12·0	14·50	1684	227	Paddle wheel; oscillating expansive.

Thornycroft's system, based on that of Ruthven's, was a great improvement on it, consequent on the experience gained in the "Waterwitch" trials; and the application of the science of hydrodynamics by Sir John Thornycroft, of which he is a past master, warranted the Admiralty to engage in some further trials in 1881, when a torpedo boat was built, 66 feet 4 inches long by 7 feet 6 inches beam, and a draught of water of 2 feet 6 inches only; the displacement was 14·5 tons and she was fitted with the Thornycroft impeller, of which a plan is shown in fig. 26. It was driven by a compound engine having cylinders 8 inches and 14·5 inches in diameter and 12 inches stroke, which developed on trial 167 I.H.P. The speed on a carefully conducted trial was, however, only 12·6 knots, which was low and a keen disappointment, seeing that a similar boat but 3 feet 4 inches shorter with 13 tons displacement ran at a rate of 17·3 knots with her engines developing 170 I.H.P.

The efficiency of the latter was not particularly high, being only 0·5, but that with the centrifugal pump was only 0·254, as estimated by Mr Barnaby.

96 MARINE PROPELLERS.

Notwithstanding this very low efficiency scarcely the last word has been said for jet propulsion, seeing that it possesses the advantage of having the propelling instrument so well protected, and the ship itself is in quite as good a condition for use as a sailing ship is for moving

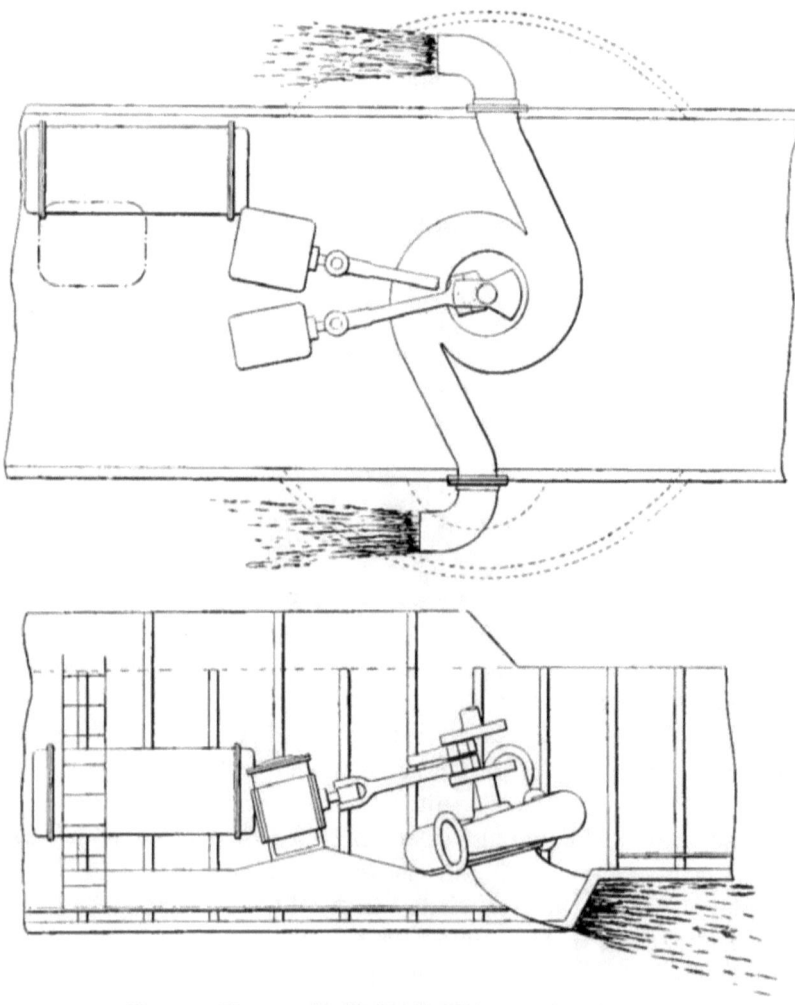

FIG. 26.—Thornycroft's Hydraulic Motor in a Torpedo Boat.

under sails. It is, therefore, eminently suited for such purposes as propelling lifeboats and similar craft that knock about in rough weather near ships and landing-places. It should be also an admirable addition to the long-voyage sailing ship to enable her to pass through calm belts, and to get into and out of harbours, docks and estuaries, inasmuch as it does not in any way detract from her ability to sail well, and high efficiency of machinery is not of first importance.

In 1888 Messrs R. & H. Green built for the National Lifeboat Institution a large lifeboat into which the hydraulic machinery of Thornycroft's make was fitted; the boat was tried, when again it was found that the efficiency was lower than that of the screw-driven boat; but, on the other hand, her *efficiency as a lifeboat* was sufficiently established to warrant them building others from time to time. The efficiency of the machinery, as measured solely by the ratio of the useful work done by it to that generated, is of small moment in a lifeboat, and is not always of prime importance in other ships. Perhaps in these days of scientific investigation accompanied by scientific phrases, a little too much importance is sometimes given to

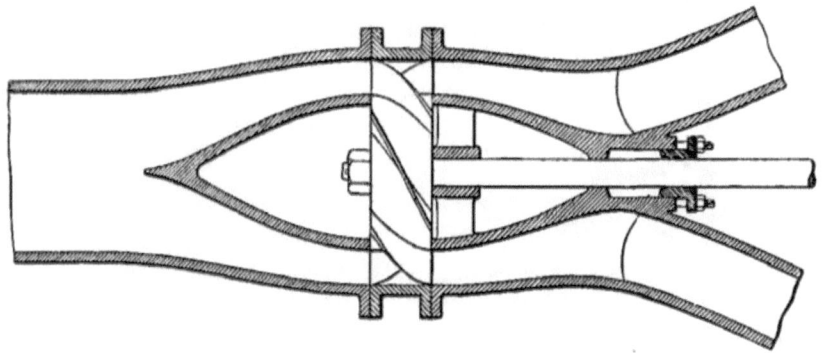

FIG. 27.—Bessemer's Hydraulic Propeller.

them, especially when they dazzle the eyes of the public so as to blind them to the real faults of a system. For example, high efficiency without reliability in continuous good running is futile. High mechanical and steam efficiency is sometimes accompanied by great liability to breakdown, and often by costly upkeep and great wear and tear. Further, high efficiency may sometimes even be purchased at such a high price that capital charges form a very serious set-off to working charges.

In the same year, 1849, that Ruthven took out his patent the late Sir Henry Bessemer proposed an arrangement of internal propeller as shown in fig. 27, which literally consists of a screw like Ericsson's, fitted in a case and with channels leading the water direct to the blade zone and from it to the stern, on either side of the wheel or screw, the case being contracted gradually to suit the channels or tubes. Inside the wheel case is an egg-shaped centre chamber tapering at each end so as to guide the water to and from the impeller; it acts as and is virtually a boss to it. The long after-end tapers with and as far as the taper of the outer case. There appears to be no record of Bessemer having tried this ingenious arrangement or done anything with it beyond the model stage, for which, however, there need be no surprise, as at that time he had not acquired the very considerable riches (whereby he could have proved this invention) that he had later on. This propeller, especially if assisted with guide blades on Thornycroft's principle, having only short channels, might have proved more successful than Ruthven's.

But after all is said, however, it remains that Bessemer's design had the same damning features as are common to all those with an internal propeller; and if it was subject to the same constructive criticism, it would cease to be an internal propeller; for it may be urged with all propriety and truth that for efficiency with such a system it is essential to keep down the velocity of flow, inasmuch as the frictional resistance of the surface of the wheel chamber and the passages to and from it is one of the chief sources of loss of energy and consequent low efficiency. Increased efficiency would be obtained by making the passages of larger sectional area, which leads to the disadvantages of taking up more room in the ship, as also of an increase in the weight of the machinery, metal passages and water required, to which may be added the loss arising from there being less space for cargo. But to counteract all these objectionable consequences, it may be urged that the inlet shall be made as close to the impeller as possible and the delivery channels be shortened to their least possible length. It is obvious, in fact, that the shorter these are, the higher must be the efficiency, and thus the maximum efficiency will be attained when their length is zero, that is, so that the impeller delivers directly to the sea. In plain language, the propeller acts best when there are no inlet or outlet passages at all.

But as some propellers, if devoid of chambers and guides, may lose in efficiency tremendously, they must be replaced by the screw or other implement which requires no external assistance to render it efficient. The fact is that few people who have strenuously advocated the adoption of the jet propeller system for general purposes can have taken the trouble to ascertain some of the physical facts and figures involved in its use on board even the ordinary ship of commerce.

To take a simple case. Suppose a passenger steamer, such as was employed a few years ago in cross-channel service, having a displacement of 1750 tons and capable of a speed of $17\frac{1}{2}$ knots on trial when the engines developed 4000 I.H.P. with an efficiency of 0·66, to be fitted up with a Ruthven installation instead of twin screws about 12 feet diameter running at 130 revolutions per minute.

With twin screws E.H.P. $= 4000 \times 0\cdot 66 = 2640$.

Now, assuming the efficiency of the hydraulic machinery, etc., to be 0·26,

$$\text{then gross I.H.P.} = 2640 \div 0\cdot 26 = 10{,}154,$$

an enormous increase in power, being 250 per cent., and such that the ship could not contain boilers enough for it.

Even if, for the sake of argument, it be supposed that by improvements, etc., an efficiency of 0·45 could be obtained, and at present that seems impossible, then

$$\text{Gross I.H.P.} = 2640 \div 0\cdot 45 = 5867,$$

a sufficiently alarming increase, and only just possible to be installed with great care.

The indicated thrust of such a ship will be

$$\frac{4000 \times 33{,}000}{2000} = 66{,}000 \text{ lbs.}$$

The actual mean thrust will be about 56,000 lbs. Then taking the usual symbols, and allowing real slip at 25 per cent., making A represent the combined area of section of passages; and assuming the speed of the ship v to be 1773 feet per minute and the mean velocity of the stream 2364, the acceleration is 591, or $\frac{591}{60} = 9\cdot 85$ feet per second.

$$\text{Then} \quad \text{thrust} = \left(\frac{A \times 2364}{60} \times \frac{591}{60} \times \frac{64}{32} \right) = 56{,}000,$$

or, $\quad A = 56{,}000 \div (39\cdot 4 \times 9\cdot 9 \times 2) = 72 \text{ square feet.}$

If there was a channel or tube at each side, the area of section would be 36 feet, or 6 feet square; rather an appalling orifice as well as an undue appropriation of space in the hold of the ship.

The velocity of flow through the passages will be at the mean rate of 2364 feet per minute, or about 23 knots. The resistance per square foot of surface will therefore be

$$\left(\frac{23^2}{10^2} \right) \times 1\cdot 25 \text{ lbs.} = 6\cdot 6 \text{ lbs.}$$

For each foot of channel the friction will be $6 \times 4 \times 6\cdot 6 = 158\cdot 4$ lbs.

The equivalent in I.H.P. of work done in overcoming this is $11\cdot 35$ I.H.P. per foot of each channel, and as the resistance of these and their bends may be equal to 300 feet of straight length, then loss from this cause alone will be 3405 I.H.P.

If so small a slip be objected to, let it be taken at 33 per cent.; then speed of flow will be $1773 \times \frac{3}{2} = 2660$ feet per minute.

The velocity of flow is now 26 knots, and hence

$$\text{friction} = \left(\frac{26}{10}\right)^2 \times 1\cdot 25, \text{ or } 8\cdot 45 \text{ lbs. per foot.}$$

Taking, however, the effect of the higher acceleration of water in reducing its quantity,

$$A = 56{,}000 \div \left(\frac{2660}{60} \times \frac{887}{60} \times 2\right) = 42\cdot 7 \text{ square feet.}$$

Each channel then will have an area of $21\cdot 4$ square feet and a side $4\cdot 6$ feet or a periphery of $18\cdot 4$ feet.

Then the frictional resistance of channel per foot $= 18\cdot 4 \times 8\cdot 45 = 155\cdot 5$ lbs.

The I.H.P. $= 155\cdot 5 \times 2660 \div 33{,}000 = 12\cdot 53$.

As before, taking the resistance as the equivalent to 300 feet of such a channel, then—

$$\text{Loss} = 12\cdot 53 \times 300 = 3759 \text{ I.H.P.}$$

To estimate the power of a jet or hydraulic installation it is necessary to know the velocity of the "feed" stream, which is the water entering the ship to supply the impeller, to take into account the velocity of the delivery stream as it issues through the nozzles or apertures, as well as the quantity of water passed through them in a second of time. If the area of transverse section of the feed channel is to that of the delivery as the velocity of flow through the latter is to that through the former, then the feed water will enter the whirl chamber of the impeller at constant velocity, due to the motion of the ship. To it is then imparted an accelerating force so that it leaves the ship at a higher velocity than that of the passing water.

If, however, the inlet is made large enough for the water to enter at the velocity v, being that of the ship, and the channel is gradually contracted, the acceleration of the water will be then gradual up to the impeller, where it receives its final impetus and may leave the whirl chamber at the high velocity V. The delivery channel may then be so tapered that its section at discharge through the side is such that the water is delivered at the velocity v. As a matter of fact, however, such physical conditions are imposed on the engineer that neither of these refinements can be carried out in practice.

The inefficiency of the hydraulic propeller is, however, due to other causes besides that of friction in passages and case; these seem to be inevitable with the system. In the case of the "Hydromotor" the active machine was practically a bad form of Newcomen engine, inasmuch as the expelling force was that of steam acting on the wooden piston and the atmospheric pressure forcing it back again with the water behind it—a single-acting condensing engine. And although the steam was supposed to exhaust to a surface condenser, no doubt a large proportion of it was condensed in the chamber. The efficiency of this arrangement would be therefore far below that of any ordinary modern steam engine.

With the centrifugal pump or other internal impeller, the losses in the chamber from various causes are huge. The wheel or impeller has an enormous surface exposed to the action of the water, so that with its high velocity the frictional resistance is most serious. The same may apply to the whirl chamber, whose sides and ends must resist the water largely and set up eddies, which are always so pernicious in their action on all such instruments. Then, too, the flow of water into the chamber, or, as it is called, the feed, which is induced by, and takes from, the power of the pump, is checked on entering it, and the whole or greater part of its energy is virtually lost. Besides these drawbacks, the whole of the water has to be lifted above the level of the sea at the expense of energy which is thus completely lost; if, however, delivery took place at or below sea flotation line, there would be then a great loss, due to dragging nozzles and fittings through the water, in fact, greater than that expended in raising it. The whirl chamber itself seems in all the experiments to have been too small for free working, and the impression gathered from examining the plans is that the water would always undergo a kind of grinding process before passing on.

The Duty and Efficiency of a Hydraulic Propeller.—If A is the combined area of section of the delivery nozzles, V the velocity in feet per second of the flow through them, and v the velocity of the ship, also in feet per second, the weight of a cubic foot of sea water 69 lbs., and gravity 32.

The quantity of water passed per second $= A \times V \times 64$ lbs.

Its mass is $\dfrac{A \times V \times 64}{32} = 2A \times V$.

The acceleration, or, as it it called, slip $= V - v$.
The reaction, or equivalent of thrust $= 2A \times V(V - v)$.
The work done is therefore $= 2A \times V(V - v)v$ ft. lbs. per second.
The energy of the "race" will be $A \times V(V - v^2)$.

These things being so, then—

The work done in propulsion = the work done + energy of race

$$= 2A \times V(V - v)v + A \times V(V - v^2).$$
$$= A \times V(V^2 - v^2).$$

The efficiency of the jet itself will be, then, $\dfrac{2AV(V-v)v}{AV(V^2 - v^2)} = \dfrac{2V}{V+v}$.

Without the modifications of channel alluded to, the energy of the feed water is dissipated and virtually lost, and is represented by $A \times V \times v^2$. This must therefore be added to the total work so as to make up the full amount expended in propulsion.

The total work is then $= 2AV(V-v)v + AV(V-v)^2 + AV \times v$.
Simplifying this it is $= A \times V^3$.

Under these conditions the efficiency of jet $= \dfrac{2A \times V(V-v)v}{A \times V^3}$

$= \dfrac{2(V-v)v}{V^2}$.

Taking the particulars of H.M.S. "Waterwitch" as given by Sir William White, viz. $A \times V = 154.7$ cubic feet, $(V - v) = 13.3$ feet; V being 29 feet and v 15.7 feet, and the I.H.P. as 760,

$$\text{Efficiency of jet} = \dfrac{2 \times 13.3 \times 15.7}{29^2} = 0.5.$$

Total efficiency of the system = useful work $\div$ I.H.P. $\times$ 550

$$= \dfrac{2 \times 154.7 \times 13.3 \times 15.7}{760 \times 550} = 0.155.$$

The resistance of the "Waterwitch" at 9.3 knots was about 4900 lbs. and the resistance H.P. $= 140$.

Efficiency in this was $140 \div 760 = 0.184$.

The efficiency of the "Viper," calculated on the same lines, was only 0·23.

The efficiency of the system as installed by Thornycroft in a torpedo boat was estimated by Mr Barnaby as follows:—

$$\text{Efficiency of engine was } 0\cdot 77$$
$$\text{\qquad,,\qquad ,, pump\qquad ,,\quad } 0\cdot 46$$
$$\text{\qquad,,\qquad ,, jet\qquad ,,\quad } 0\cdot 71$$

Total efficiency of system $= 0\cdot 77 \times 0\cdot 46 \times 0\cdot 71 = 0\cdot 2514$.

It is unfortunate for us that no progessive trials were made with these ships. Even the old "half-boiler" trial, if made with the "Waterwitch," would have given some good information on which to make further investigations, and perhaps have given some indications of the road to follow to avoid the heavy losses.

CHAPTER VIII.

THE SCREW PROPELLER: LEADING FEATURES AND CHARACTERISTICS; THRUST AND EFFICIENCY.

A screw propeller is an instrument consisting of a central hub or boss secured on the end of a shaft projecting through the end of the ship, to which boss are affixed two or more wings or blades, each of which is shaped and formed like the others and set angularly equidistant from one another with their centre lines at right angles, or nearly so with the axis of the shaft. The face or surface of the blades acting on the water is helical; that is, it is formed of a part of one or parts of more than one helix.

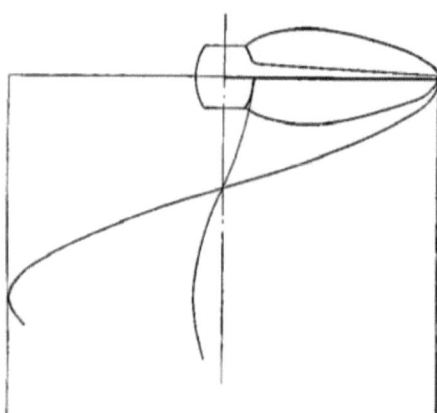

FIG. 28.—Screw Blade on a True Helix.

The helix is a surface developed by the revolution of a line about one end, which end travels or advances lineally on an axis at right angles to it as it revolves. (See fig. 28.) A true helix is thus formed when the angular movement and the lineal advance are synchronous; that is, if L is the length of the one complete convolution of the helix and $n \times l = L$, then an axial advance of l and an angular movement of $360° \div n$ are made in the same time.

The pitch of a helix is the length of the axis or the distance travelled when a complete revolution of the describing line has been made; or it may be defined as the distance between two parallel planes at right angles to the axis, which thus cut off a complete convolution of a helix.

The angle of the helix at a distance r from the axis is one whose tangent = pitch $\div 2\pi r$.

The surface of the helix of radius R, pitch P, is

$$R \times \sqrt{P^2 + \pi^2 R^2} = R\sqrt{P^2 + 10R^2}.$$

Common screw was the name given to the screws first used by F. P. Smith and for a long time after in H.M. service. The boss was cylindrical and comparatively small in diameter, the acting surface was a part of a helix cut by parallel planes a fraction of the pitch apart, latterly about one-eighth. Taking the length as l, then—

Surface of each blade of a common screw $= \dfrac{l}{P} \times R\sqrt{P^2 + 10R^2}.$

As the common screw had usually two blades and sometimes more, each was a portion of a separate helical convolution cut off by the parallel planes.

Developed Surface.—Such screws naturally had blades with very wide tips, and the actual acting surface which was fan-shaped was said to be the developed surface when laid out on a plane surface.

Projected surface of a screw is that made by projecting or throwing the shadow of the blades on one of the transverse planes; its area is often taken as the measure of the ability of the screw to make thrust. At one time there was a disposition to consider this surface only, but it is not convenient, as it varies with the alteration in pitch, whereas that of the acting surface of the blade itself remains the same.

Diameter of a screw is that of the circle described by the outermost point of a blade in making a revolution.

True screw is the name now applied to the screw whose whole acting surface of blade is part of one helix only. That is, the pitch is uniform throughout.

Variable pitch screws are those whose blade surfaces are helical, but made up of portions of various helices, as the pitch varies from one part to another.

Bennet Woodcroft claimed special virtues for the screw, the blades of which had a coarser pitch at the following part than at the leading edge (see fig. 29). He thought that the water would be thus gradually set in motion sternwards and *leave* the screw at the highest velocity instead of starting it suddenly and traversing the blade at the same speed. With a slow revolution screw there would be in reality this theoretical gain, but with the screws of to-day, whose speed at the tip is 60 to 70 knots, it is scarcely to be expected that such a refinement as Woodcroft patented would make any difference. Still, there are plenty of engineers of the mercantile marine who seriously claim that such propellers do give better results in the steamers under their observation, while others seem equally convinced that there is really more virtue in a propeller whose pitch does not vary transversely, but which does increase from the boss to the tips; others, again, even fancy a propeller with a coarser pitch at the boss than at the tip. In the naval service a true screw is the rule now, and it is that generally used in the mercantile service.

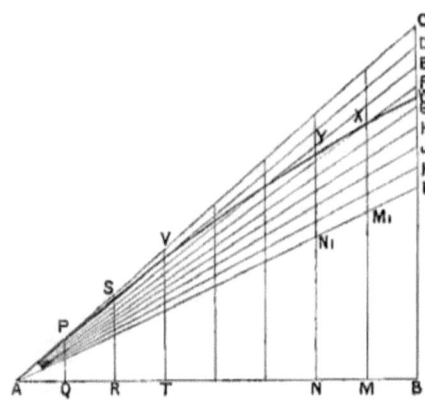

FIG. 29.—Increasing Pitch Helix (Woodcroft).

Woodcroft's original screw was of one convolution, so that the pitch gradually increased from 10 feet at the fore to, say, 12 feet at the after or delivery end, AYW.

Fig. 29 shows the line AYW, due to a pitch varying from BL to BC, and is constructed as follows:

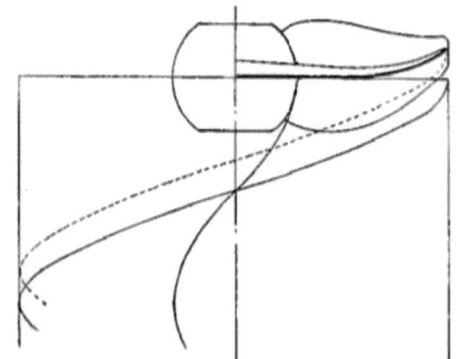

FIG. 30.—Screw Blade Curved Forward, Griffiths' patent.

The base-line AB represents the circumference of the circle whose diameter d is that of the propeller at any point required; therefore $AB = \pi d$.

BC is perpendicular to AB and is the pitch at delivery, while BL is the pitch at entry edge of blade.

Divide LC into the same number of parts (say 8) as AB. Then BK is the pitch at ⅛ the distance from the leading edge, and AK is the angle made by the blade at that part, and so on for BJ, BH, etc. to BC.

Draw QP, RS, TV, etc. perpendicular to AB; AP is then the eighth part of AC.

Draw PS parallel with AD cutting RS at S_1, S_1 parallel with AE, V_1 cutting TV at V_1, etc. etc., and XW parallel with AL.

The polygonal line APVXW can be resolved into a curve, and if the figure is then wrapped around on a cylinder of circumference equal AB, APVXW is the trace of the helix of increasing pitch required.

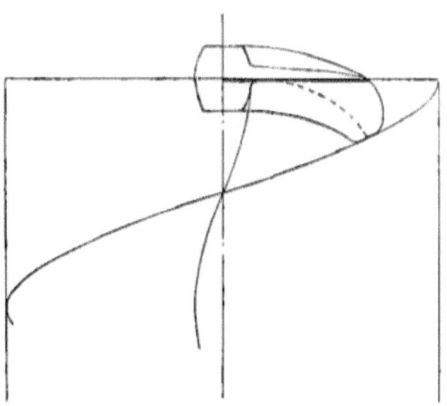

Fig. 31.—Screw Blade thrown back by making centre line Spiral on the Bed.

But Woodcroft and others, as screws got shorter, made the leading half of the blade of one pitch, the following of another, and faired one into the other with the curvature shown in fig. 56.

Griffiths and others since his time have shown a preference for a blade of which the acting surface is developed by the revolution of a curved line instead of a straight one; in Griffiths' case the generating line would be straight for a half to two-thirds of its length and then bent towards the bow. (See fig. 30.) Very many engineers prefer the surface to be developed by a straight line at less than a right angle with the axis, so that the surface is a portion of what is called by mechanics a "V thread." (See fig. 32.)

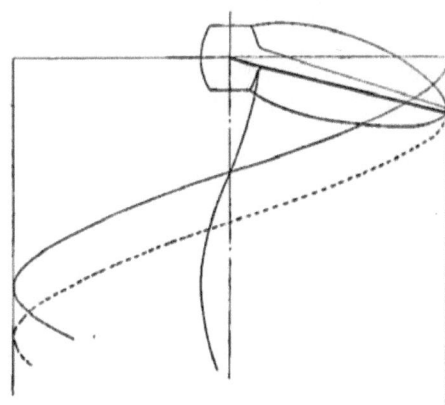

Fig. 32.—Screw Blade thrown back by Coning the Bed.

In all these cases every portion of the blade is that of a true helix, as there is no variation in pitch. Some makers of screw propellers who like to have the blade ends thrown away from the ship, do so on a true and common helix by curving the centre of the blade on it so that the tip is lower than even the end of the boss (fig. 31).

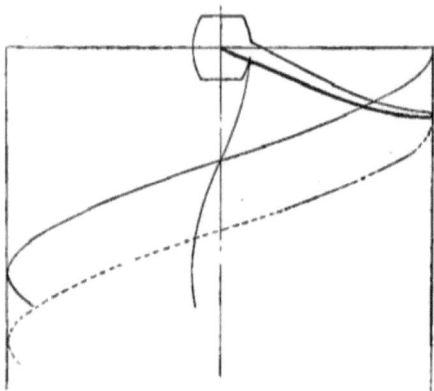

FIG. 33.—Screw Blade Curved Back.

Others, again, make a bent blade like the Griffiths, but make the generating line at less than 90 degrees with the axis. (See figs. 32, 33.)

Pitch ratio is the ratio of the pitch to the diameter of the screw, $P \div D$; the exigencies of to-day to suit steam turbine motors require the pitch ratio to be often less than 1·0. Experience with screws driven by reciprocators has shown in the past that a low-pitch ratio was not satisfactory, and generally resulted in a low efficiency of screw; it was also often accompanied by, if not the cause of, negative slip. On the other hand, screws with pitch ratios of 1·2 to 1·5 invariably give satisfactory results accompanied by moderate positive slip.

Surface ratio is the ratio of the area of the acting surface of a screw to that of a circle of the diameter of the screw.

With two-bladed screws	Common	Surface ratio	0·275 to 0·320
,, two ,,	Griffiths	,, ,,	0·260 to 0·290
,, three ,,	Common	,, ,,	0·350 to 0·400
,, three ,,	Griffiths	,, ,,	0·335 to 0·380
,, three ,,	Admy. leaf	,, ,,	0·270 to 0·310
,, three ,,	Circular	,, ,,	0·370 to 0·400
,, three ,,	Broad tips	,, ,,	0·420 to 0·440
,, four ,,	Common	,, ,,	0·388 to 0·450
,, four ,,	Admy. leaf	,, ,,	0·330 to 0·380
,, four ,,	Circular	,, ,,	0·500 to 0·520
,, four ,,	Broad tips	,, ,,	0·530 to 0·550
,, four ,,	{ Mercantile ordy. Square tip }	,,	0·290 to 0·400

THE SCREW PROPELLER.

In these pages it is proposed to use always the following:—

P the pitch and D the diameter of the propeller in feet.
V as the velocity of the propeller in feet per second.
v ,, ,, ,, ship ,, ,,
S ,, ,, ,, propeller in knots.
s ,, ,, ,, ship ,,
R as the revolutions of the propeller per minute.

Velocity of propeller is the pitch multiplied by the number of revolutions.

$$V = \frac{P \times R}{60} \text{ feet per second, or}$$

$$S = \frac{P \times R}{101 \cdot 3} \text{ knots.}$$

Resistance of the ship, tow rope, is that due only to the resistance of the ship to her passage through water, as is the case when being towed without a propeller by another ship at a considerable distance from it. Such resistance is measured by the tension of the tow rope, as was done by Dr William Froude in his experiments with H.M.S. "Greyhound." He found that the resistance of the "Greyhound" was 0·6 ton at 4 knots, 1·4 tons at 6 knots, 2·5 tons at 8 knots, 4·7 tons at 10 knots; at 12 knots 9·0 tons, which was excessive, as this ship was designed for 10 knots only. A much larger ship, the "Merkara," which had a resistance of one ton at 4 knots, had only the 9 tons at 12 knots. The resistance per square foot of wetted skin was 1·396 lbs. at 10 knots with the "Greyhound." With modern ships at 10 knots the resistance is 1 lb. per square foot of immersed skin when fresh painted to 1·4 lb. in ordinary clean condition. (*Vide* Table V., Chapter III.)

Augment of resistance is the increased resistance caused, first of all, by the increase in velocity of the water past the hull by the suction of the propeller ; and, secondly, by the diminution in hydraulic head or pressure at the stern due to the action of the screw. Screws of large diameter and fine pitch have a special tendency to aggravate this loss of "head."

Total resistance of the ship is the resultant of the resisting forces thus created and made active; it is these that the propeller thrust has to overcome, and thrust is equal to them when the ship is moving at uniform speed.

Resistance of propeller, if of no pitch, is that due to the skin friction on moving through the water and to the resistance due to the bluntness of the edge and the form of the body forced through the water. The energy absorbed in overcoming this is all lost, and amounts to a large fraction of the total energy imparted to the screw by the engines; for at a velocity through the water of 10 knots each square foot of surface like that of a screw propeller exerts a resistance of $1\frac{1}{4}$ lb. (in ordinary bronze blades or those merely painted). Highly polished blades and those having fresh enamel paint will offer less—probably only 1 lb. per square foot.

The back of the blade may not set up so great an amount of friction, but it will not be materially less, taking into account its shape, etc., so that it is usual, in calculating frictional resistance of screws, to take twice the acting surface and assume that the mean resistance is based on an allowance of $1\frac{1}{4}$ lb. at 10 knots at any part of it in its passage at that. The resistance will vary, of course, as the square of the velocity, so that those propellers whose rate of motion near the tips is 60 knots will resist at the rate of 45 lbs. per foot at the tips, or, taking back and front, 90 lbs. per foot of acting surface. The other resistances named will vary roughly with the number of blades, but inversely as the breadth of blade, and in everyday practice with good screws may be taken at 5 per cent. of the total skin resistance for each blade. That is, if there are four blades, the total resistance is 1·2 multiplied by frictional resistance.

Frictional resistance of a screw blade may be found by the following simple methods:—Fig. 34 shows the outline of the developed surface of half a blade whose figure is symmetrical about GB. The propeller is moving at a uniform rate of revolution so that BC represents the velocity *through the water* at the tips to a convenient scale.

That is, the velocity per revolution at B and at any intermediate point is

$$v_1 = \sqrt{\text{pitch}^2 + (\pi d)^2},$$

d being the diameter at any point taken.

If BC, etc., GK, represents on a convenient scale the velocities at B, etc., G. A curve drawn through C, etc., K will permit of the velocity being ascertained at any intermediate points by taking the intercept between BG and CK at these points. The resistance per square foot may be calculated at three or four points by the rule $\gamma = 1\cdot 25 \left(\dfrac{S}{10}\right)^2$ lbs. and a curve GD set up in the same way so that intercepts will give the resistance at any intermediate points.

Now, taking narrow strips of the blades at three or four stations and multiplying by the resistance at these stations and doubling the result to allow for the blade backs, a curve HE is obtained so that intercepts again give the resistance at various stations, and the area is the measure of the total resistance of one blade.

Proceed, then, to multiply the resistance of the strips as obtained

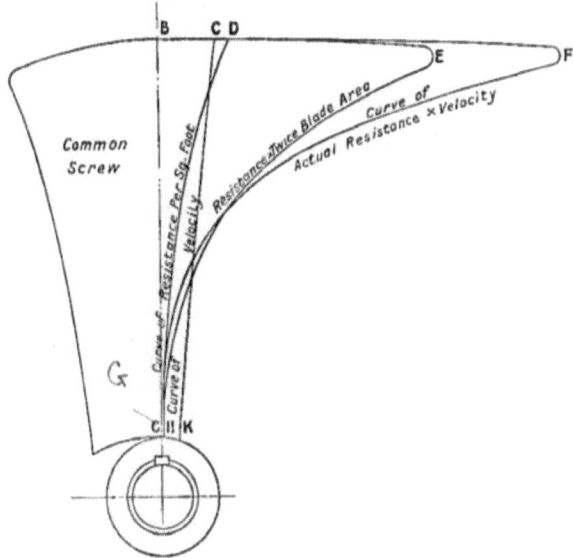

FIG. 34.—Showing Curves of Friction, Resistance, etc., of a common Screw Blade.

above by the space moved through by them in a minute, and the work absorbed in turning that blade is measured by making a curve HF by means of a few of the ordinates so found as before.

Intercepts between HF and GB give the work absorbed in moving those strips through the water, and the area GBFH represents the total power in ft. lbs. absorbed in turning that blade through the water.

Dividing it by 33,000, the horse-power required to overcome h is obtained.

Figure 34 represents the equivalent resistance of two of the four blades of H.M.S. "Amazon," and fig. 35 is that of one of the two of the Griffiths screw which replaced it and gave so much better results.

The ill effect of the broad tip is seen at a glance, as are also the losses arising from excessive diameter, for, by taking six inches off each tip, the resistance is in both cases very much reduced, especially so in the case of the four-bladed screw. Froude found the efficiency of the "Greyhound's" machinery to be exceedingly low, and attributed it chiefly to engine resistance, whereas it was largely due to the absurdly large diameter of the screw, it being 12·33 feet diameter with 52 square feet of surface; whereas the "Rattler," of similar size and power, had a screw 10 feet diameter with only 22·8 square feet of surface, which elaborate experiments years before had shown to be sufficient. Moreover, the "Rattler" had a speed coefficient of 224 against that of 142 of the "Greyhound," which ought to have opened the eyes of the authorities in 1865.

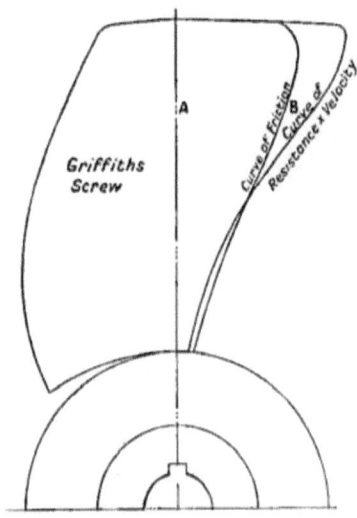

Fig. 35.—Showing Curves of Friction, etc., of Griffiths Screw.

With the high speed of revolution necessary for the efficient working of turbine motors, as also for the speed of revolution possible with modern reciprocators, especially the enclosed variety with automatically forced lubrication, propellers of small diameter are absolutely necessary for safe running, while to prevent cavitation the blade area must be relatively large. Hence we find the modern propeller is gradually getting nearer and nearer in width of blade to our old friend the common screw of sixty years ago, and differs from it now chiefly in its having nicely rounded corners instead of the rigidly square ones of our grandfathers' time.

Fig. 36 shows one blade of H.M.S. "Rattler" of 1845; the dotted line is that of a blade of a modern turbine motor steamer. Now, although the difference in blade is small to look at, the action when at work is very different. Those corners of the old screws caused violent vibration at high speeds; but when they were cut away there was a very marked improvement.

Frictional resistance of a screw propeller may be calculated with a close approximation to the truth by taking the velocity at the tip and the total area of acting surface, using multipliers in both cases deduced from the close calculation of it with screws of different types.

Let S be the velocity of the blade tips in knots per hour.

Let R be the revolutions per minute.

Let D be the diameter in feet.

Let P be the pitch of screw in feet.

Let A be the area of acting surface in square feet.

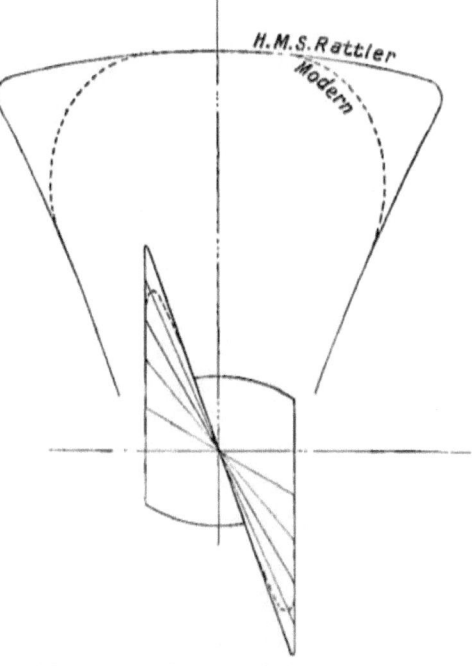

Fig. 36.—Modern High Revolution Screw compared with that of H.M.S. "Rattler," 1845.

The resistance of a square foot is assumed to be $1\frac{1}{4}$ lb. at 10 knots.

$$S = \sqrt{P^2+(\pi D)^2} \times R \times 60 \div 6080 = R \times \sqrt{P^2+(\pi D)^2} \div 101\cdot 3.$$

Resistance per square foot $= 1\cdot 25 \left(\dfrac{S}{10}\right)^2$ lbs.

Resistance of screw $= 2A \times 1\cdot 25 \left(\dfrac{S}{10}\right)^2 \times f$ lbs.

For a common screw	$f = 0\cdot 634$
For a fantail-shape screw	$f = 0\cdot 581$
For a parallel blade	$f = 0\cdot 550$
For an oval	$f = 0\cdot 520$
For a leaf shape	$f = 0\cdot 450$
For a Griffiths	$f = 0\cdot 350$

The horse-power absorbed in overcoming the frictional resistance may be found now by multiplying the resistance by the space in feet moved through in a minute and dividing by 33,000.

The mean space moved through by the blade surface from tip to boss of an ordinary propeller $= 0.7 \times$ distance moved through by the tip.

$$\text{Hence mean space} = \frac{S \times 0.7 \times 6080}{60} = 70.9S.$$

$$\text{Then I.H.P. expended} = 2A \times 1.25 \left(\frac{S}{10}\right)^2 \times f \times 70.9S \div 33{,}000$$

$$= \frac{A \times S^3 \times f}{18{,}612}.$$

Example.—A screw 12 feet diameter, 15 feet pitch, has 42 square feet of surface and moves at 130 revolutions per minute (three leaf blades).

$$\text{Here } S = \sqrt{225 + 1440} \times 130 \div 101.3 = \frac{40.8 \times 130}{101.3} = 52.4.$$

$$\text{Frictional resistance screw H.P.} = \frac{42 \times 52.4^3 \times 0.515}{18{,}612} = 166.8.$$

Edge resistance here will be 3×5 per cent. or 15 per cent. of $166.8 = 25$ H.P.

Then total resistance of screw $= 166.8 + 25 = 191.8$ H.P.

The thrust of the screw is the resultant of all the pressures on the screw acting in a direction parallel to its axis and applied through the screw shafts to the thrust block. There is no method of calculating this with any degree of certainty from purely theoretical considerations; therefore for H.M. ships and other ships of high speed and great importance, the thrust and other characteristics of any proposed screw are ascertained by experiments with models in a tank fitted with special apparatus for towing and accurate recording instruments. This is, of course, both an expensive and tedious method, and a luxury at present only enjoyed by a few; moreover, such experiments, while being very interesting, are not necessarily conclusive as to what may be expected from the full size screws.

Indicated thrust was a term introduced by the late Dr William Froude as a means of expressing a value for the thrust of a screw at various speeds by which its performance could be analysed and compared with that of other screws; and, although he did not claim that the thrust so estimated was the actual thrust, it was not difficult to deduce from it something approximate to it. It is clear, however, that even this claim must be taken with reserve, seeing how misleading may be the results of such calculations with screws of abnormal

THE SCREW PROPELLER.

proportions, as will be shown later on, unless they are treated with caution.

Froude's rule was as follows:—

$$\text{Indicated thrust} = \frac{\text{I.H.P.} \times 33,000}{V} \text{ in pounds.}$$

For example, suppose a ship to travel at 15 knots with the I.H.P. 3000, and the apparent slip of screw $12\frac{1}{2}$ per cent.;

Here $\quad v = 15 \times 6080 \div 60 = 1520$ feet per minute.

$$\text{Slip} = V - v = \frac{V}{8},$$

then
$$V = \frac{8}{7} v = \frac{8}{7} \times 1520 = 1737 \text{ feet per minute.}$$

$$\text{Indicated thrust} = \frac{3000 \times 33,000}{1737} = 57,000 \text{ lbs.}$$

Practically this means that the *power* delivered by the screw is $\frac{7}{8}$ of the I.H.P. With the 20 per cent. slip of the older engines Froude's suggestion did give a good approximation to the real one; but with the machinery of to-day with 20 per cent. slip it would not be nearly so correct.

Froude also very properly pointed out the directions in which the power generated in the engine is expended, and drew up a sketch balance-sheet as a model for engineers to follow in making up their accounts.

First of all, the engine itself requires a certain amount of the gross power it develops to be absorbed, as it were, in overcoming its own resistance in the way of friction, and for losses due to the inertia of the moving parts.

Secondly, the working of the air, feed, bilge and other pumps make further inroads on the I.H.P.

Thirdly, the resistance of the shafting and thrust requires another portion.

The balance is the power transmitted to and put into the screw, and may amount to 90 per cent. of the I.H.P. or even more, with good modern engines having only air pumps to drive and all the working parts carefully fitted and thoroughly lubricated. In the older ships with circulating, feed and bilge pumps, in addition to the air pumps, worked by the main engines and having surface condensers and working parts well made throughout, the balance was about 85 per cent.;

while with horizontal jet condensing engines and the bearings and guides such that there was always a tendency to run hot, the balance was often as low as 70 to 80 per cent.

If N.H.P. be the gross power imparted to the propeller, then—

Efficiency of the engines and shafting $= $ N.H.P. $\div$ I.H.P.

Of this power a considerable portion is required to overcome the resistance of the propeller itself, as already shown, and T.H.P., the balance, should be employed in making "thrust"; but it may be that some more power is being wasted in disturbing and making eddies in the water and dissipating its energy in other ways than that of "projecting a stream of water in the direction opposite to that of the ship's motion."

The net balance, in any case, is employed in making thrust, and the power may be called *Thrust Horse-power* or T.H.P. Then:—

Efficiency of the propeller $=$ T.H.P. $\div$ N.H.P.

Total efficiency of machinery and screw $=$ T.H.P. $\div$ I.H.P.

But, as already shown, the screw when working is always the cause of an increase to the ship's resistance; in some cases that increment is a most serious one; hence if the tow-rope resistance of the ship is R and the H.P. corresponding to it is T_R.H.P., then:—

True efficiency of machinery, propeller, and the ship $=$ T_R. H.P. $\div$ I.H.P.

The resistance of the ship can generally be estimated from the information obtained from trials at low powers when the wave-making is not serious enough to be taken into account, and when there could be no cavitation or other disturbing causes at the propellers, or by assuming that each foot of wetted skin has a resistance of 1 lb. to $1\frac{1}{4}$ lb. at 10 knots.

General efficiency of propeller is a thing hardly attainable in theory, inasmuch as it is evident that the screw propeller whose pitch and surface are the best for a speed of 20 knots cannot be the propeller of maximum efficiency at 15 knots. It is therefore a very debatable point whether it is better to design the screw to suit the maximum contract trial speed, or for the service speed in the case of mail and passenger steamers, or for the maximum speed consistent with naval service conditions in H.M. Navy for warships.

The Actual Thrust of a Screw Propeller.—In the early days of screw propulsion attempts were made to measure it by means of a dynamometer applied to the screw-shaft end so as to take the whole thrust. With the shaft geared to the engine shaft, as was then customary, this method was easily adaptable, but the "readings" were such as to render any results deduced from them to be always taken with reserve, and they were generally open to extreme doubt; it is therefore not surprising to find the records published by the Admiralty of trials of H.M. ships with the dynamometer quite inconsistent one with another; and only those taken on board H.M.S. "Rattler," when it may be presumed the instrument was new and carefully calibrated, are worthy of careful analysis.

Since those days other attempts have been made to measure the thrust of a screw propeller of a size beyond the model stage—notably those of Mr Yarrow with a torpedo boat, as published by him in a paper read before the Institute of Naval Architects. His experiments were ingeniously devised and most carefully carried out, so that all disturbing elements were minimised and most of them eliminated.

Mr Yarrow's experiments made in 1883 with a 60-ton torpedo boat are very interesting and instructive. He carried out a series of trials at speeds varying from 9 to 15 knots. (See p. 237.)

(1) Propelled by her own engines, from which the indicated horse-power was noted as I.H.P.

(2) The thrusts in pounds and the horse-power corresponding were also carefully measured and noted as T.H.P.

(3) After removing the propeller the boat was towed by another and larger boat, so as to be as free as possible from any disturbance due to the latter; the tension on the tow-rope was carefully measured and noted and the horse-power deduced, denominated $T_R.H.P.$

It was found that the maximum value of $T_R.H.P. \div I.H.P.$—that is, the maximum efficiency of the whole—was 0·672 at about 11 knots while at 9 knots and 15 knots it was 0·63.

The maximum efficiency of engines and propeller as shown by T.H.P. ÷ I.H.P. was also at about 11 knots and the value 0·852; at 15 knots it was only 0·733, while at 9 knots it was 0·800.

The maximum efficiency of the propeller as taken by the ratio $T_R.H.P.$ to T.H.P. was 0·860 at 15 knots, while at 9 knots it was 0·786 and at 11 knots 0·790.

Mr Yarrow expressed some doubt as to the perfect accuracy of the I.H.P., and judging from some observations made with quick-running engines that have come under the author's notice, there is reason to suspect that Mr Yarrow was right, and that the power is quite 10 per cent. below what it should have been had the instruments been fitted direct to the cylinders. Making this addition, and deducting a fair allowance for engine friction, etc., it would appear that the losses at the propeller itself were 1·5 H.P. at 10 knots, 4·1 at 11 knots, and as much as 70 at 15 knots.

Taking these figures, and calling the net horse-power delivered to the screw N.H.P., then T.H.P. ÷ N.H.P. is a maximum at about 10 knots, and is 0·759, while at 15 knots it was only 0·61.

Under these conditions the screw losses at 10 knots were only 5 per cent. of the N.H.P., while at 15 knots they amounted to 29 per cent. The augment of resistance at 10 knots was 20 per cent of the N.H.P., and at 15 knots only 10 per cent. of the N.H.P.

Mr R. E. Froude, whose researches and extensive experience in screw propeller experiments not only entitle him to a most respectful hearing, but to the gratitude of every marine engineer and naval architect for the most useful and invaluable information he has given them without stint ever since he succeeded to the place so long occupied by his honoured father Dr William Froude, has propounded a formula for calculating the actual thrust of a propeller and given the multipliers deduced from carefully made model experiments, so that it can be applied to any screw. He states:—

The analysis, then, of the series of thrust and efficiency curves yielded by the experiments on the individual model screws was in the first instance based on the following simple formula for thrust in terms of revolutions per minute:—

$$T = aR^2 - bR \qquad . \qquad . \qquad . \qquad (1)$$

Where T = thrust; R = revolutions per minute; a = a coefficient depending on dimensions, etc., of screw; b, one depending on the speed and pitch.

This formula embodies the following idea, which under certain ideal conditions would be theoretically correct. In any screw revolving in still water at various rotary speeds without axial advance the thrust will, of course, be proportional to the square of the rotary speed. This fact is expressed by the term aR^2 of the formula, represented by the ordinates of a parabola ABCD. If now we suppose the screw, while still revolving as before at various rotary speeds, to have a definite forward axial linear speed of advance V as well, there will, of course, then be a certain definite rotary speed (revolutions per min. $= R_0$, say), at which the thrust will be zero, below which it will be negative, and above which it will be positive, but of decreased amount. And the formula expresses this decrease of thrust in terms of revolutions per minute by the negative term bR, represented by the ordinates to the straight line ACE, cutting the parabola at C, namely, at R_0, the revolutions of zero thrust. Thus the straight line ACE becomes in effect the zero line for the curve CD, regarded as the thrust curve for the speed of advance V; and, similarly, the same parabola ABCD may be made to furnish the thrust curve for the same screw at any speed, by drawing accordingly the sloping line ACE which represents the term $(-bR)$ of the formula. This formula expresses the thrust curves of experiment, in general, remarkably well; since, therefore, a new reduction was in any case necessary, this formula was unquestionably the right basis for at any rate the primary analysis of the results.

Thrust.—Continuing the study of the formula, it will be seen that if, as is most convenient for such analysis, and as has been done in this case, we take as a conventional measure of the pitch the travel per revolution at the revolutions per minute R_0 (of zero thrust) so that

$$P = \frac{V}{R_0},$$

and, again, observing that for zero thrust at revolutions $= R_0$ we must have $aR_0^2 = bR_0$, we can eliminate the coefficient b in terms of a, V, and P. And, assuming the coefficient a to have been correctly obtained for a screw of specific design and unit diameter (diameter $= D = 1$), we obtain from equation (1) above the following two alternative equations for thrust:—

$$T = \frac{a}{p^2} D^2 V^2 \frac{S}{(1-S)^2} \quad . \quad . \quad . \quad . \quad (2)$$

$$T = aD^4R^2S, \quad . \quad . \quad . \quad . \quad (3)$$

where

$$p = \frac{P}{D},$$

or pitch ratio, and $S =$ slip-ratio as ordinarily reckoned, viz. $(R - R_0) \div R$. The former of these, expressing thrust in terms of speed and slip-ratio, is perhaps the most intelligible; while the latter is often more convenient for computation.

To enable the thrust for given speed and revolutions to be

calculated by either of these formulæ for a propeller of any given design and dimensions, it only needed to determine the coefficient a as affected by difference of design, the principal elements in which may be taken to consist in (i.) pitch ratio; (ii.) type, and blade width proportion.

It was found that the effects of these two principal elements might be taken as independent of one another, and that, as regards (i.) a might be most correctly taken as proportional to $p(p+21)$. As regards (ii.), the value $\dfrac{a}{p(p+21)}$, which, as just seen, is constant for varying pitch-ratio, was taken as the expression for the "blade-factor" B; the purpose of which is to denote what may be called the thrust capacity of the propeller, as dependent on type, *i.e.* whether three-blade elliptical, three-blade wide-tip, or four-blade elliptical; and within each of these types, on width proportion of blade. The value of this blade factor B as obtained from the experiments, and as dependent on these variants, has been indicated by the ordinates of three curves respectively proper to the three types just mentioned, on an abscissæ scale representing blade width proportion, as indicated by "disc area ratio," namely, ratio of total blade area to disc area.

At the same time, for a final test of the formula of equation (1) as accurately expressing the variation of thrust with revolutions, the thrust values of all the individual propellers, at a series of slip-ratio values, were carefully compared with thrust values calculated by formula; and on this information the thrust formula was corrected by multiplying the right-hand side by $1 \cdot 02\,(1 - \cdot 08\,S)$. Making this correction, and also substituting for a its value in terms of the blade factor B just referred to, equation (2) becomes the final thrust formula, as follows:—

$$T = D^2 V^2 \times B \frac{p+21}{p} \times \frac{1 \cdot 02\,S(1 - \cdot 08\,S)}{(1-S)^2} \qquad (4)$$

To facilitate calculations, a curve was computed expressing the last factor (involving S only) as an ordinate ($=y$) to a base ($=x$) expressing revolutions and pitch relatively to speed, as indicative of the slip ratio S. This curve is commonly called the "xy" curve. Conveniently for ship screw calculations, the numerical coefficients used in the computation of this curve were chosen for expressing, not thrust, but "thrust horse-power" (or T.H.P.) $=$ H; speed $=$ V, in knots; revolutions = R, in hundreds; diameter = D, in feet. We thus get as the expressions for x and y as follows:—

$$x = \frac{RpD}{V}\left[= \frac{1\cdot 0133}{1-S} \right] \qquad (5)$$

$$y = \frac{p}{B(p+21)} \times \frac{H}{D^2 V^3}\left[= \cdot 0032162 \frac{S(1-\cdot 08\,S)}{(1-S)^2} \right] \qquad (6)$$

TABLE VIII.—B VALUES.

Disc area ratio .	·30	·35	·40	·45	·50	·55	·60	·65	·70	·75	·80
Four blades, elliptical	·0978	·1020	·1050	·1070	·1085	·1100	·1112	·1124	·1135	·1147	·1157
Three blades, wide tip	·1045	·1097	·1126	·1148	·1166	·1182	·1195	·1207	·1218	·1230	·1242
Four blades, elliptical	·1040	·1106	·1159	·1197	·1227	·1249	·1268	·1282	·1294	·1306	·1318

Dr W. Froude's analysis was proposed by him as a means whereby the efficiency of a screw could be determined, and its performance compared with other screws was arrived at as follows:— He assumed that at each revolution the mean pressure on the pistons multiplied by twice the stroke was equal to the thrust multiplied by the pitch of the screw.

Let A be the area of L.P. cylinder or cylinders in square inches, L the length of stroke, P the pitch in feet, and p the referred mean pressure.

$$\text{Thrust} \times P = p \times A \times 2L, \text{ or thrust} = \frac{p \times A \times 2L}{P}.$$

Multiply both numerator and denominator by R, the number of revolutions per minute, then

$$\text{thrust} = \frac{p \times A \times 2L \times R}{P \times R}.$$

Now $p \times A \times 2L \times R = \text{I.H.P.} \times 33,000$, and, substituting, then

$$\text{thrust} = \frac{\text{I.H.P.} \times 33,000}{P \times R}.$$

This he called indicated thrust, and by calculating it for the various speeds taken on progressive trial, and using the results as ordinates to the speeds as abscissæ, a curve drawn through their ends shows the indicated thrust at all speeds (fig. 37, p. 122), and the trend of the curve indicates whether there is a falling off or abnormal increase in thrust as the speed increases. If the latter, it is evident the screw is doing its duty, but the ship is not responding; if, on the other hand, the curve falls away the screw is too small for the power put into it, or is using the power in some other way than that of producing thrust.

Froude's curve of indicated thrust (fig. 37) is constructed as follows:—

On the base line OX take points B, C, and D corresponding to the three different speeds of progressive speed trial. From the data given, calculate the indicated thrust by the following rule:—

$$\text{Indicated thrust} = \frac{\text{I.H.P.} \times 33{,}000}{\text{pitch} \times \text{revs. per min.}}$$

Draw BE, CF, and DG perpendicular to OX and proportionate to

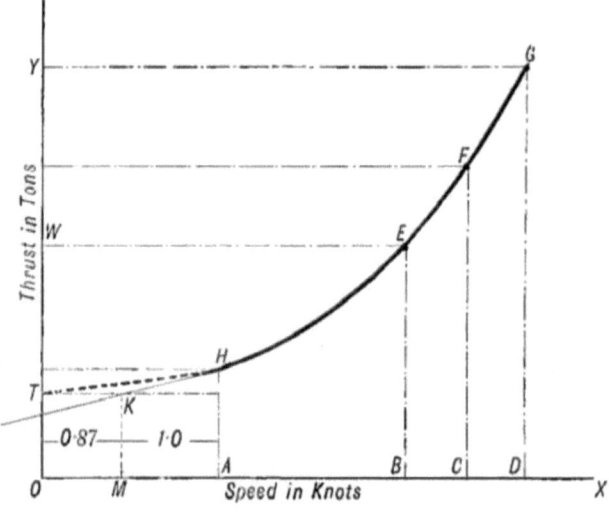

FIG. 37.

the indicated thrust thus calculated, on a convenient scale. Now take A so that OA represents the speed at the slowest rate of steaming observable, calculate the indicated thrust as before, and erect AH as representing the amount to the same scale as the others, and through HEGF draw a curve and continue the same if possible, so as to cut OY at T. As it is generally, in practice, impossible to get so slow a speed as to enable the prolongation of the curve HEGF to be made with accuracy, Froude suggested that the point T might be obtained by taking a point M between OA so that—

$$\frac{\text{OM}}{\text{OA}} = \frac{0\cdot 87}{1\cdot 0}.$$

Draw MK perpendicular to OX, and touching the tangent drawn through H at the point K. Through K draw KT parallel to OA, cutting OY at T, which is then the point through which the curve should pass.

If such a curve is produced it will not pass through the origin but cut the ordinate some way above it; this is the measure of the thrust at *no speed*, and it was thought by Froude to indicate "the equivalent friction of the engine due to the working load."

Froude differentiated and laid down that "when decomposed into its constituent parts, indicated thrust is resolved into several elements which must be enumerated" as—

(1) *The useful thrust* or ship's true resistance.

(2) *The augment of resistance*, due to the diminution which the action of the propeller creates in the pressure of the water against the stern end of the ship.

(3) The equivalent of *the friction of the screw blades* in their edgeway motion through the water.

(4) The equivalent of the *friction due to the deadweight of the working parts, piston packings*, and the like, which constitute the initial or slow-speed friction of the engine.

(5) The equivalent of *friction of the engines, due to the working load*.

(6) The equivalent of *air pump and feed pump duty*.

It is probable that (1), (3), and (4) are all very nearly proportional to the useful thrust; (6) is probably nearly proportional to the square of the revolutions; (5) is constant at all speeds—that is, the *power* absorbed in engine friction varies with the revolutions.

A practical rule for estimating thrust of a screw has been devised by the author as the result of analysing a series of trials made at various times of actual ships by various engineers with such accuracy, both of observation and calculation, as to command respect and permit of their use in an estimation of the actual thrust. Mr R. E. Froude, like others, is of opinion that thrust varies with the acting surface; and, generally speaking, if it is understood that there is no other variable function, such as pitch or diameter changed, this is fairly true; but it is still not absolutely true, for the effect of adding a square foot of surface near the tips will be very different from what would follow from making the same addition near to the roots of the blades. The broad statement is therefore by itself absolutely wrong and misleading, for it is manifest that a screw

10 feet diameter having a surface of 40 square feet would produce a very different thrust from that given by a screw having the same surface but 13 feet diameter. Moreover, experience has shown that with two blades the total surface may be smaller than with a propeller of the same diameter and pitch, but with three or four blades

Professor Coterill has shown in a very interesting paper read before the members of the I.N.A. that it is only a common screw, that is, one with very broad tips, that can make a complete column of water, and then only at its outer part.

The author, for these and other reasons, therefore, prefers to take diameter and square root of blade area multiplied together as one of the governing functions instead of area simply. He also finds that within the limits of practice thrust varies inversely with the pitch ratio.

Then, if A is the aggregate area of active blade surface in square feet, D the diameter in feet, V the speed of the screw in feet per second, P_r the pitch ratio, and G the ratio of the distance of the centre of gravity of the blade face *from the boss* to half diameter of screw, the rule is:—

$$\text{Thrust in pounds} = \frac{D \times \sqrt{A} \times V^2}{P_r} \times G.$$

This will give, approximately, the mean thrust of any ordinary screw, that is, of any screw whose shape or proportions are not abnormal. In practice the following values for G may be taken when it is not convenient to calculate them:—

Griffiths blade, broad	$G = 0.36$
,, ,, narrow	$G = 0.33$
Oval and leaf-shaped round tip	$G = 0.40$
Circular blade	$G = 0.42$
,, broad tipped blade	$G = 0.45$–0.50
Mercantile square-tipped blade	$G = 0.42$

Examples.—(1.) To find the thrust from the screw of a torpedo boat whose diameter is 6·5 feet, pitch 8 feet, revolutions 350, area of blades 12 square feet, leaf-shape.

$$\text{Thrust} = \frac{6 \cdot 5 \times \sqrt{12} \times 48^2}{1 \cdot 23} \times 0 \cdot 4 = 16{,}830 \text{ lbs.}$$

(2.) To find the thrust on each of the twin screw of an Atlantic liner whose diameter is 18 feet, pitch 25 feet, blade area 81 square feet and Griffiths broad type. Revolutions 90 per minute.

$$\text{Thrust} = \frac{18 \times \sqrt{81} \times 37\cdot5^2}{1\cdot39} \times 0\cdot36 = 58{,}990 \text{ lbs.}$$

(3.) To find the thrust on the propeller of a cargo steamer whose diameter is 20 feet, pitch 24 feet, surface 110 square feet, leaf-shape. Revolutions 60 per minute.

$$\text{Thrust} = \frac{20 \times \sqrt{110} \times 24^2}{1\cdot20} \times 0\cdot40 = 40{,}270 \text{ lbs.}$$

Mean pitch of a screw is usually meant to be the mean of all the pitches as measured at a series of positions on the acting face of the blades, adding them together and dividing by the number. This, of course, is an arithmetical mean only, and is convenient for using when calculating the slip. But it is by no means a true measure of the active or effective pitch of any screw, and far from being so when the difference of pitch is large and varies on a rule or method such as that devised by Woodcroft, Atherton, and others. Woodcroft and others down to Sir J. Thornycroft, who made screws with the pitch of the following half of the blades considerably greater than that of the leading edge, claim that the water is set in motion gently and the total acceleration given to it only completed at the following edge. If this is so, the pitch of the following portion of the blades is the effective pitch and the mean of it taken as the mean effective pitch. In making calculations for such screws as Woodcroft's some such method must be adopted for arriving at a measure of the really effective pitch.

Again, in the case of the screw, which varies in pitch from the boss to the tip, so that the greatest pitch is farthest from the axis, no arithmetical mean will give the true indication of the effective pitch of such a screw. From observation, the pitch at the broadest part or that just beyond the middle of the blade seemed to be a fair measure of the capacity of the screw; and so long as it is a true screw circumferentially the pitch may be measured on positions on that circle and the arithmetical mean taken as the acting pitch.

A further complication arises when a screw is made to vary in pitch from edge to edge and at the same time to vary from root to tip. Who can possibly say what is the effective pitch of such a screw, and how can any arithmetic mean possibly be taken as an expression of it?

So many of the ships showing negative slip had screws with varying pitch in the old days, while the true pitch common screw seldom gave it, that there was always a strong suspicion that wrong estimation of effective or acting pitch was why the slip came out negative; and even now there is reason to look on it with great suspicion—especially on the curved sections, which may set up an acting or active *water surface* apart from the metallic surface as shown in fig. 15.

Loss by water "slip."—There is another source of loss with the screw in common with all propellers, and that is from the frictional resistance of the water as accelerated in its passage through the surrounding water. It will be seen by reference to fig. 16 that the water in flowing to the propeller from the orifice would be subject to resistance in the channel through which it flows, and a certain amount of the power expended on suction would be in this way used up and wasted. Further, it will be seen in the same figure that in passing the screw and beyond it, there is more resistance with its consequent waste.

Now in the ordinary screw steamer similar passages, etc., are formed in the still water as shown in fig. 17, through which the stream to and from the propeller flows. But a portion of these passages is formed by the skin of the ship itself, so that the friction on it is used up in retarding the motion of the ship. This amounts to a considerable quantity in paddle steamers, and will not be small in those screw steamers having three and four propellers. In the case of the icebreaker "Ermack" with four screws, it is said her speed is better with the bow screw going "*astern*" than "ahead," because of the influence of its stream on the ship's bow and skin.

The efficiency of a propeller is measured on the same principle as all other instruments, and tested in the same way as everything else must be sooner or later in commercial life, viz. by comparing the *useful* work done by the propeller with the cost, that is, the power imparted to it while occupied in doing it.

The screw, however, differs from most other instruments, inasmuch as it may have two quite different values assigned to it for its output, depending on whether it is looked on as a "pusher," when the gross thrust is the force exerted by it, or simply as a propeller, in which latter case the resistance of the ship overcome by it is the measure of its useful work.

The power imparted to the screw or paddle wheel is that

developed by the engine, less the amount required to move itself and its appurtenances. The older vertical engines driving paddle wheels or geared to a screw shaft had two large so-called "air pumps," the capacity of each of which was generally one-eighth the capacity of the cylinder. The horizontal engine for driving screw propellers had two double acting pumps each about one-twelfth the capacity of the cylinder. These pumps had to deal with large quantities of water as well as produce and maintain a vacuum of 26 inches; consequently a large portion of the I.H.P. developed was absorbed in this duty alone, and it probably varied as the square of the revolutions, as would also that for the other pumps, of which each engine had a pair of bilge pumps and a pair of feed pumps of considerable size to allow for "blowing down" the boilers.

It may be taken as fairly correct that the N.H.P. of the old jet condensing engines when well made was about 80 per cent. at full speed. In the newer engines with surface condensers, and the circulating of water done by centrifugal pumps, the efficiency at full speed may have been 85 to $87\frac{1}{2}$ per cent.; in newer engines still, having three cranks and only the air pump worked by the main engine, 90 to 93 per cent. is not an extravagant estimate for the best made vertical ones, judging by some experiments made on engines running with and without the propellers.

The ordinary friction of a marine engine practically varies with the revolutions, for although at very slow speeds the friction per revolution increases, the rate of revolution is then much less than generally obtains when dealing with propeller problems.

In a general way the friction of an engine is proportioned to the size of its cylinders, that is, to the nominal horse-power; for this purpose Nom. H.P. may be estimated by multiplying the diameter by the stroke of the low-pressure cylinder (both in inches) and dividing by 15 for a compound, 13 for a triple, 10·5 for a quadruple engine.

A fair allowance for internal resistance of a modern engine is ·006 per Nom. H.P. per revolution.

Example.—What is the frictional or internal resistance of a triple engine having cylinders 20 inches, 32 inches, and 52 inches diameter and 36 inches stroke, when running at 150 revolutions?

Nom. H.P. $= 52 \times 36 \div 13 = 144$.

Friction $= ·006 \times 150 \times 144 = 129·6$ I.H.P.

At this speed the I.H.P. would be about 1800.

The efficiency is therefore 92·7 per cent.

The net horse-power thus imparted to the screw is absorbed in turning it round at so many revolutions per minute, besides causing it to overcome its own resistance, frictional and otherwise, in passing through the water; it imparts acceleration to a mass of water; it may also in so doing set up side currents, that is, a whirling action due to the obliquity of blade and friction of surface.

The imparting of motion to the water axially sets up a thrust along the axis as already explained, which thrust may, or may not, be wholly employed usefully.

Augmented resistance is produced more or less by every screw propeller used in the stern of a ship, inasmuch as it must take away some of the pressure on the stern due to the "head" of water. The larger the screw the greater must be the augment, and with a very large screw it is enormous, as may be seen by referring to the "Archer's" trials, wherein a screw 12·5 feet diameter was at first used when one 8 feet in diameter would have been sufficient; and also severe when placed close to the ship, as in the case of the "Dauntless."

The useful effect of a screw is, therefore, the gross thrust due to the acceleration of the column of water less the augmented resistance set up by it.

In comparing the performances of ships and propellers, much may be done with simple means, and that without doing violence to scientific truth, as by assuming that the tow-rope resistance of a ship varies as the square of the speed and as the wetted skin, as indeed it does in all well-formed ships for speed, rather below that of full speed; that for copper-sheathed ships, such as the old naval frigates and corvettes, an allowance of $1\frac{1}{4}$ lbs. per square foot at 10 knots may be taken, in a general way, as fair, although with *clean new copper* carefully nailed, $1\frac{1}{8}$ would be sufficient; that the early iron ships with such paints as were then used probably set up a resistance of 1·1 to 1·125 lbs. per square foot; and that with the modern spirit enamel paints, freshly put on, an allowance of 1 lb. is enough.

That is the rule for—

$$\text{Tow-rope resistance} = \text{WS} \times \left(\frac{S}{10}\right)^2 \times f.$$

S is the speed in knots.
$f = 1·0$ for the best enamel paints.
$f = 1·1$ for the older anti-fouling compositions.
$f = 1·25$ for coppered ships.

Tow rope horse-power $= WS \times \left(\dfrac{S}{10}\right)^2 \times f \times$ feet moved through by the ship per minute

$$= WS \times S^3 \times f \div 32{,}560.$$

This is the power required to move the ship through the water without producing heavy waves, and is, of course, the ideal condition. It may therefore be taken as the standard for comparison when dealing with ships and their machinery and propellers. Hence—

General efficiency . . .	= Tr. H.P. ÷ I.H.P.
Propeller efficiency . .	= Pr. H.P. ÷ Net H.P.
Engine efficiency . . .	= Net H.P. ÷ I.H.P.

With such means as these it is comparatively easy to analyse the results of any trial in the better way than merely noting the speed coefficients; and when, in addition to a full-power trial, there is one at a lower power as in the old Navy days of "half-boilers in use," there is a capital means of checking the results observed at the higher speeds.

The more recent trials in the Navy always include one at 10 knots, which is of course now by comparison a very low-powered one. Ships as now designed can move at this speed without producing waves of sensible magnitude; hence the resistance, then, is practically skin friction only.

CHAPTER IX.

VARIOUS FORMS OF SCREW PROPELLER.

SMITH'S and Woodcroft's original screws consisted of a complete convolution of one helix around a boss having a bearing at each end. Both inventors soon dropped this form and resorted to one consisting of two half-convolutions, getting thereby with the same surface better results generally, and with a marked reduction of vibration in particular. Advancing further, both shortened their screws until the length was only about one-eighth of the pitch, with results still more gratifying and fully justifying the changes. For many years all screws were made of about that length, and to-day there is no departure from that practice worth noting.

Robert Griffiths, who commenced to devote himself to a study of the screw propeller at that time, was apparently the first to perceive that the chief defect of the common screw was in having such a broad tip; that is, so much of its surface was relatively remote from the axis of revolution; he first of all proposed to make the blades as shown in Plate V., fig. 37, but eventually he came in 1860 to the design now so well known, and an example of which is seen in fig. 38, p. 131, which was the propeller of H.M.S. "Galatea," a very large frigate and famous in her day.

Other engineers, however, had suggested blade forms at about the same time that differed largely from the common screw in respect to the distribution of acting surface and worked fairly well, but whereas their proposals have all been dropped and probably most of them forgotten, that of Griffiths practically remains to-day; inasmuch as the fundamental form of nearly every screw now in use has all the leading features of his.

Another development which had a fascination for and was suggested by more than one engineer in those early days of screw propulsion was that of cutting the corners away and making the blades of practically the same width from root to tip. The desire to have a propeller that was masked by the stern-post when the blades were in the vertical position so as to cause practically no obstruction when the ship was under sail no doubt led to, and prompted the

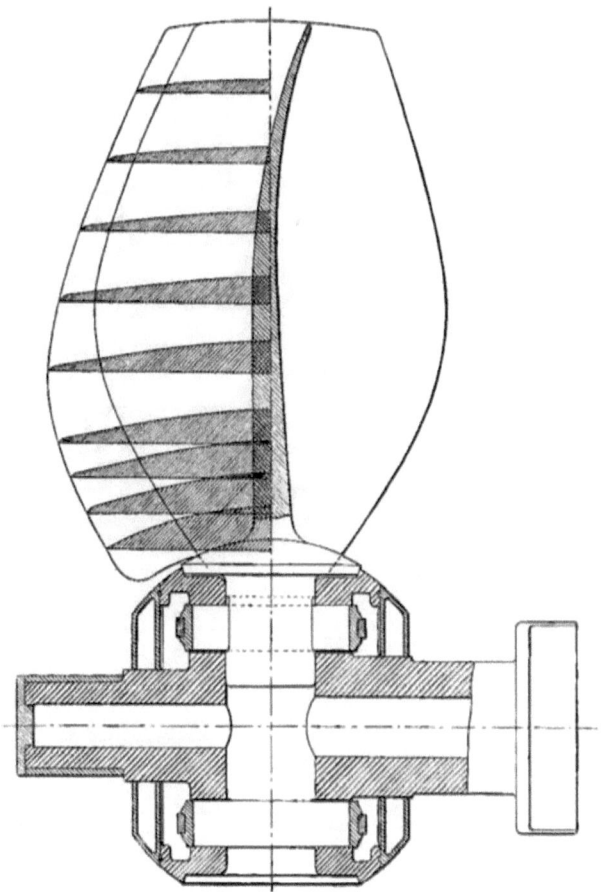

FIG. 38.—Propeller on Griffiths' Patent (1860) Adjustable Blades.

adoption of, this form of blade; and with a view further to minimise this, the temptation to make the blades narrow was great; the result was that such propellers generally had insufficient surface and so suffered from what is now called "cavitation," and was evidenced by excessive slip.

Mangin Screw.—To overcome this defect, Mangin, a French engineer, constructed a propeller which was virtually a pair of narrow two-bladed propellers placed in line one behind the other, as shown in fig. 39. He likewise followed Woodcroft's idea by making the blades with an increasing pitch, the leading quarter of each blade being finer than the following three quarters; sometimes, however, the following pair of blades were of greater pitch than the leading pair. These screws were fitted first to H.M.S. "Flying Fish" in 1854, and throughly tested; later on quite a considerable number of ships in H.M. Navy, as the result of these experiments, were fitted with

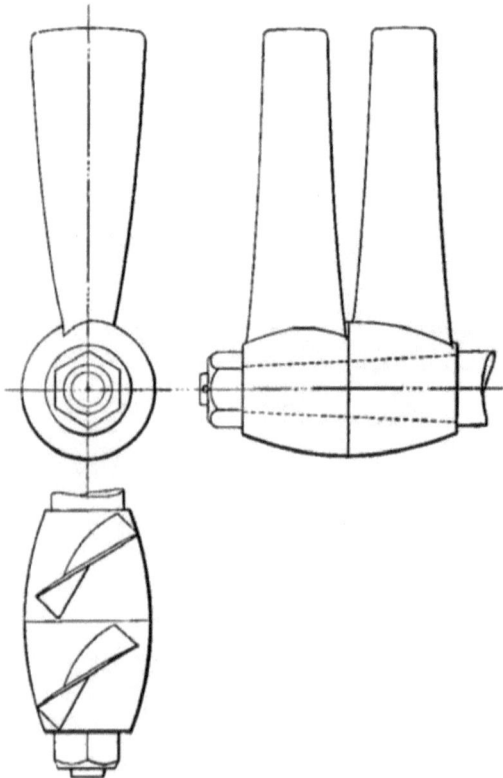

FIG. 39.—Mangin's Double Screw.

these screws, say from 1860 to 1870, most of them having twin screws. A few merchant ships were likewise supplied with them; they were apparently considered to be successful at that time, but judging from the results of trials (see p. 133) it will be concluded that it must have been rather on account of the smallness of their obstruction than to any superiority in propelling power that such a verdict was arrived at. With the reduction in the sails on steamships the Mangin propeller disappeared from service, and its use has never been revived.

It may, however, be that some of the virtue it possessed was due to the same causes that made for the success of the two and three propellers of small diameter that circumstances compelled Mr Parsons and others to fit to the screw shafts to utilise the power transmitted through them.

Some further interesting experiments were made with the Mangin screw in H.M.S. "Flying Fish" in 1857, when that ship was tried with each half screw separately and then with both halves fixed with the blades at an angle of 21·5° and then with them parallel, one ahead of the other; the results may be seen below. As might have been anticipated, the slip with the after half only was enormous, although the efficiency, as shown by the speed coefficients, compared favourably with that of the complete screw.

TABLE IX.—TRIALS OF MANGIN'S PROPELLERS ON H.M.S. "FLYING FISH," 1857.

Conditions of Screw.	35 Two Halves 21·5° Apart.	36 Two Halves Set in Line.	37 Forward Half Only.	38 After Half Only.	39 Common Screw.
Diameter of screw . ft.	13·2	13·2	13·2	13·2	13·17
Pitch of screw . . ,,	20·1 F / 25·8 A	20·1 F / 25·8 A	20·1	25·8	19·95
Number of blades .	4	4	2	2	2
Area of acting surface, sq. ft.	48·8	48·8	23·6	25·2	56·4
Pitch ratio . .	1·52 F / 1·96 A	1·52 F / 1·96 A	1·52	1·96	1·52
Surface ratio . .	0·360	0·360	0·173	0·185	0·415
Revolutions per minute .	76·75	70·75	101·0	89·0	79·0
Slip . . . per cent.	28·31 F / 44·17 A	21·34 F / 38·75 A	45·49	51·25	25·83
Speed of ship . knots	10·794	11·025	10·908	11·038	11·536
Indicated horse-power .	1093	1093	1270	1177	1052
,, thrust . lbs.	20,640	22,170	20,650	16,750	22,00
$D^{2/3} \times S^3 \div I.H.P.$.	117·8	125·5	104·7	116·9	148·9

It will be noted also that in the case of the Mangin screw the slip was excessive, but it seems singular that it was least with the blades in line—that is, one behind the other. It is also a little curious that the speed with the forward screw only should have been practically the same as when that screw followed the other by 21·5°, and that with the after half the same is obtained with both in line.

Improved Common Screw.—Emulated by the success of the Griffiths screw, and with the intention of keeping from the scrap-heap so many of the common screws as were then in existence, the naval authorities commenced to reduce their broad tips by cutting away the leading corner so as to approximate in surface and breadth of tip to the Griffiths. A marked improvement in vibration was manifest in all cases, and where the original screw had had an excessive surface the speed for I.H.P. was better. Further attempts were made in the same direction by cutting away the following corners, so that the blade was now an elongated hexagon. The results in this case were disappointing so far as the speed was concerned, even although there was a further diminution in vibration. A good instance of this treatment may be seen in the case of H.M.S. "Doris," p. 213. The area, after both corners had been cut off, was insufficient.

Hirsch's Screw.—Although patents continued to be taken out for improvements in screw propellers, none of them are worth noticing until 1860, when Herman Hirsch prescribed and patented a form of screw whose leading features are set out in patent 2930 (p. 29). Later on, in 1866, he took out another patent for improvements on his former; the new screw was as shown by fig. 40.

From 1870 to 1875 the shipping world was from time to time excited by the reports circulated of the wonderful improvements made in speed and coal consumption, especially the former, of ships that had been fitted with the Hirsch screw in place of one of the ordinary type. There is little or no reason to doubt the accuracy of the statements made at that time, but an analysis of them has since made manifest that in most cases of success the Hirsch screw had replaced one of very bad design and proportions; and further, that where the Hirsch screw had seemed to fail, it was only so because it had been in competition with a highly efficient screw that was already doing all that could be expected under the conditions; that is, when the old screw was driving the ship at the highest economic speed possible with the form of the ship, the Hirsch could do little or no more. And if an improvement in speed was obtained by it in some other ships, it was generally on account of the higher power developed by the engines with the new screw; the coal consumption was then not always so satisfactory. The fact is, that Hirsch knew so much of the right principles on which to design a propeller that he always supplied a fairly good screw and one suited to each special ship submitted to him; whereas, at that time, the majority of screw propellers were designed by rule of thumb in a happy-go-lucky way that sometimes produced good results and more often bad ones; and as there were no progressive trials at that time, or other methods of analysing and differentiating the results of trials, it was never known with certainty to what to attribute apparent failures, or rather *comparative* failure, for half a knot more or less short on trial speed was not generally reckoned to be serious enough to justify further trials, or even troublesome investigations. How very far unfit screws

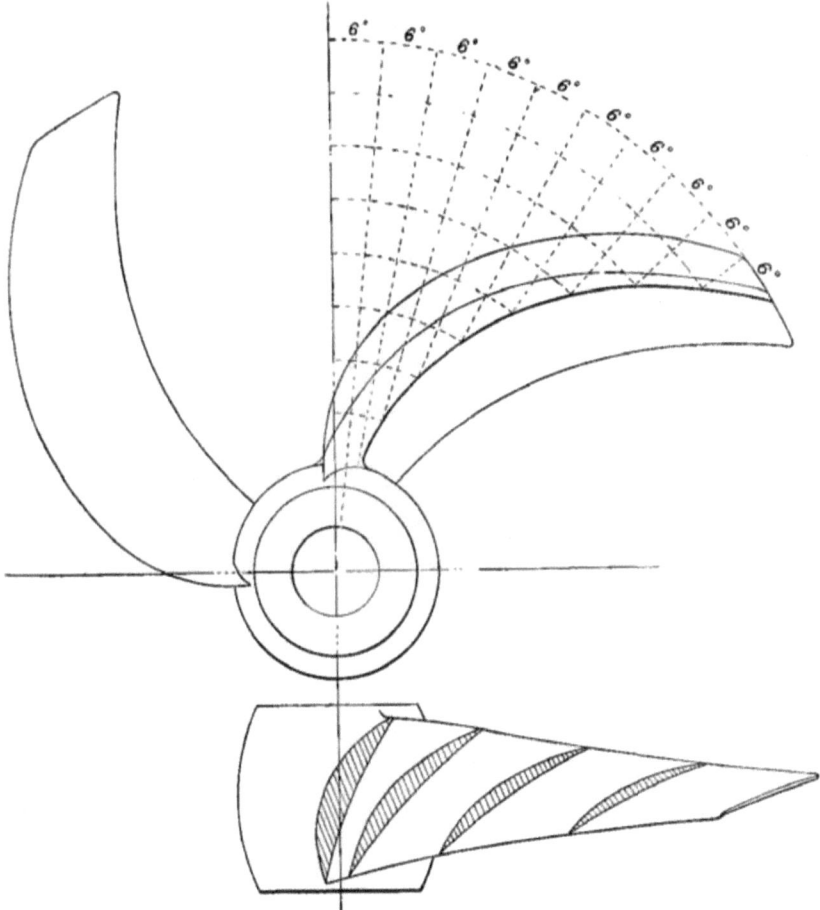

Fig. 40.—H. Hirsch's Screw of 60°. Patented 1866.

can contribute to speed failure was never appreciated until the trials of the "Iris" were made and analysed.

The White Star s.s. "Adriatic" was fitted with a Hirsch screw in the place of the ordinary four-bladed one. Ten voyages across the Atlantic were made with each screw. The average time with the old screw was 18 days 9 hours 18 minutes; with the Hirsch screw it was

136   MARINE PROPELLERS.

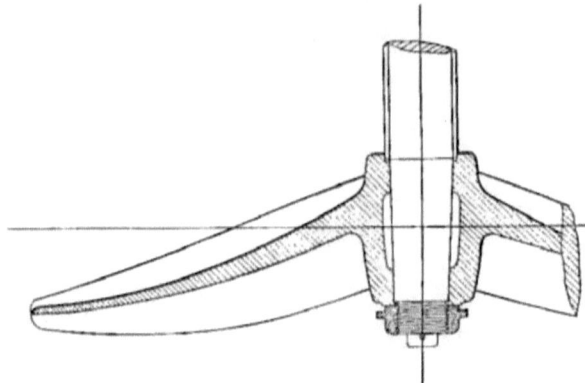

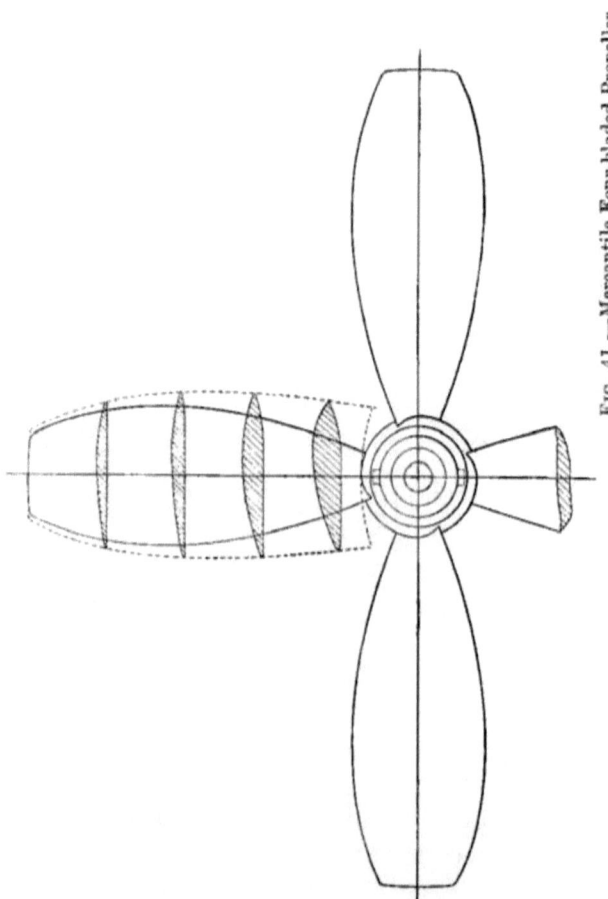

FIG. 41.—Mercantile Four-bladed Propeller.

only 17 days 5 hours 1 minute. The Hirsch in that ship was stated to be a very efficient screw when going "astern," and no doubt it would be.

Hirsch's screws were said to be more free from vibration than other screws, and instances were given which seemed to prove this. No doubt with some of the older ones the Hirsch screw would compare favourably under any circumstances, but we know now that vibration is not always due to the screw, and that a comparatively small change in number of revolutions per minute will effect a radical change in it. Consequently, if the old screw was revolving at a rate which synchronised with the ship's period of vibration while the Hirsch was faster or slower, the change for the better would be observable.

It also follows that if the power wasted in vibrating the ship was,

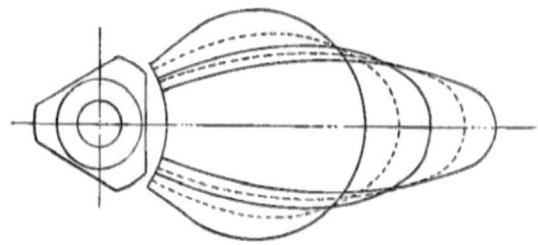

FIG. 42.—Oval Blades of Equal Surface for Different Diameters.

as it could be, applied to propulsion, the coal consumption or speed would thereby be improved.

The following facts were at the time vouched for by reliable witnesses:—

		knots.		knots.
The "Louise" (German Navy), Griffiths screw,	13·37 ;	Hirsch,	14·07
S.S. "Herder" (Mercantile),	,,	,,	11·48	,,	12·38
S.S. "L'Isere,"	,,	,,	7·985	,,	8·99
S.S. "Louisane,"	,,	,,	10·05	,,	11·10
S.S. "Pereire,"	,,	,,	14·459	,,	15·459
S.S. "Conrad,"	,,	,,	9·595	,,	10·65

Consumption of coal per 100 miles.

S.S. "L'Isere,"	Griffiths,	4·55 tons ;	Hirsch,	4·03 tons.
S.S. "Conrad,"	,,	11·18	,,	,,	10·28	,,
S.S. "Louisane,"	,,	23·32	,,	,,	20·77	,,

The ordinary mercantile cast iron screw for cargo steamers is made as shown in fig. 41, with the boss and blades cast in one piece. The centre line of the blade is sometimes as shown, and sometimes the blades are "thrown aft" by the other means described in Chapter VIII. Sometimes the screw is made with the face square with the shaft, with only a slight curvature near the tip.

Fig. 42 shows the contour of the blades of four screws differing

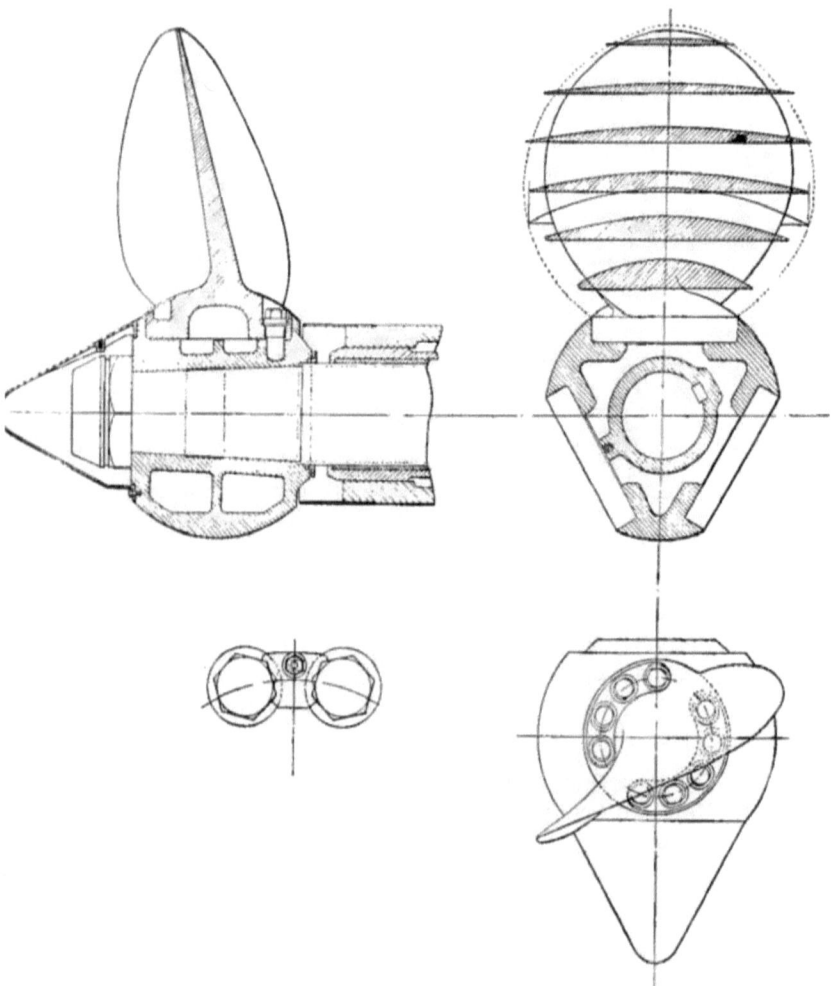

FIG. 43.—Modern Bronze Naval Screw.

largely in diameter but having the same area of acting blade surface, the blade shapes in each case being elliptical. The surface taken for purposes of comparison is 36 square feet, and the diameters range from 9 feet to 12 feet. The pitch of each is 15 feet, and the calculation is based on the revolutions being also the same, viz. 100 per minute. The speed of the ship would be in each case about thirteen and a half knots.

The following is calculated by means of the formula given for *Thrust* on page 124 and for blade friction on page 113.

TABLE X.

Diameter of screw . . . . ft.	9·0	10·0	11·0	12·0
Pitch of screw . . . . ,,	15·0	15·0	15·0	15·0
Pitch ratio	1·667	1·500	1·364	1·250
Surface ratio	0·566	0·458	0·379	0·319
Speed at tip . . . knots	31·8	34·5	37·4	40·3
Thrust as calculated . . lbs.	10125	12500	15125	18000
Friction ,, . . ,,	545	643	756	875
Thrust horse-power	421·6	520·9	620·2	725·0
Friction ,, ,, . .	46·5	59·5	75·9	94·7
Edge resistance, do . .	9·3	11·9	17·2	18·9
Total horse-power	477·4	592·3	711·3	838·6
Friction and edge H.P ÷ Thrust H.P.	0·132	0·137	0·147	0·157
,, ,, ÷ Total H.P.	0·117	0·121	0·128	0·135

These oval forms are those in general use to-day in the Navy and express steamer service. In the case of turbine-driven screws the blades are fuller still, and in some cases so much so that they resemble a square with the corners well rounded off.

Fig. 43 shows in detail the screw propellers as now used in the Navy with reciprocating fast running engines; the shape of the blade is a common one, but in some cases it is even fuller at the tips. The blades usually have oval holes in the flanges, so that their position may be shifted and the pitch increased or decreased as may at any time be desired. By fitting "cod" pieces in the ends of the holes touching the bolts, the blades cannot shift.

CHAPTER X.

THE NUMBER AND POSITION OF SCREWS.

The single screw continued in vogue for many reasons till the 'sixties of the nineteenth century, although several inventors and pioneers had claimed to fit more than one screw, and others insisted on two screws. Trevithick in 1815 claimed for his patent screw propeller that "it may revolve at the head or the stern of the vessel; or one or more such worms may work on each side of the vessel." Other inventors also suggested fitting more than one screw to a ship, and the general idea at one time was to substitute the spiral for the paddle wheel on each side.

F. P. Smith both in his model ship and the "Archimedes" fitted a single screw in the deadwood aft; the Admiralty did the same thing when building their first screw ship, the "Rattler"; and continued the practice until 1866, when they built H.M.S. "Penelope."

In the mercantile marine the single screw has continued to be the rule down to the present day for these reasons. The hull of the ship is somewhat cheaper when designed for a single screw, and no guards, etc., are necessary on each quarter to protect the screw. There is, moreover, one engine less to look after, which is important in a tramp or coasting steamer, with its very limited staff of engineers. With a single engine one engineer on watch can look after the engines and other machinery. With twin screws, even of comparatively small size, it is practically difficult, if not impossible, to get one man to keep watch single-handed, although with a single screw of the same I.H.P. and appurtenances he would willingly do so. More floor space is occupied by twin-screw engines, and likewise more light and air space overhead, which often means cutting into the 'tween decks. The engine shafting and propellers are more costly and sometimes rather heavier, power for power, than single-screw ones. There are double the number of working parts to attend to, to consume oil, and to require adjusting in port. Finally, there are likewise double the number of parts liable to fracture or hindrance.

On the other hand, the liability to break down, that is, to stoppage, is less; about a half that of the single screw. If one screw is so damaged as to be useless, the ship can still steam at a respectable speed with the other; and if the rudder or steering gear is disabled, the ship can be navigated by manipulating the engines.

Moreover, in the case of large high-powered ships the engines may each be of so much more moderate a size as to become less costly to make, and certainly less so to repair or overhaul. Take the case of the s.s. "City of Berlin" of 1874 with one screw and a low-pressure cylinder 120 inches diameter and 66 inches stroke 5200 I.H.P.; or that of the "Etruria," single-screw ship 14,500 I.H.P., built in 1884, which had two low-pressure cylinders 105 inches diameter and 72 inches stroke; while the "Celtic," with her large power (13,100 I.H.P.), has two low-pressure cylinders, one to each engine 98 inches diameter and 63 inches stroke; and H.M.S. "London," of 15,500 I.H.P., has a low-pressure cylinder to each engine only 84 inches diameter and 51 inches stroke.

It is, of course, obvious that a twin-screw ship is safer and handier than, and can be manœuvred in a way impossible with, the single-screw ship; and also equally plain that when the draught of water of a ship is such that a single screw of sufficient diameter, etc., for the engine power cannot have proper immersion, two or more screws must be employed. In fact, the first twin-screw ships were so designed because of the shallowness of the water they had often to enter and navigate. Rennie's ships of 1853-4 (see page 28) were required for service on the Nile when it was low; the twin-screw ships built in the early 'sixties of the nineteenth century were for blockade running; they had to be of large power and of so light a draught that they could get in and out of the Southern ports through channels that were not navigable by the Federal war ships blockading them. After the American Civil War such of these vessels as survived were employed in the Cross-Channel and North Sea express and other services, where their light draught and speed was of advantage in competing with paddle steamers.

The British Admiralty built a very considerable number of twin-screw ships from 1865 to 1870, mostly small cruisers and gunboats,[1] but the "Captain," of 7672 tons, the "Audacious" class of four ships,

[1] *Vide* Chap. XVIII.

5560 tons, were quite large ships and were the means of fully demonstrating as well as convincing the naval authorities generally of the superiority of the system for naval purposes. For cruising purposes it was at first intended to use one screw only, the other being dragged through the water without revolving. Two of the ships were tried in this way with the results set out below, which show that at about $10\frac{3}{4}$ knots on H.M.S. "Invincible" it takes about 427 more I.H.P. to drag the screw than to revolve it, while on the "Vanguard" it took 953 more I.H.P. to drag one screw at 11·36 knots. (See Table XI.)

TABLE XI.—TRIALS OF CERTAIN TWIN-SCREW SHIPS WITH ONE AND BOTH SCREWS RUNNING. LENGTH 280 FEET, BEAM 54 FEET, DRAUGHT OF WATER 21 FEET, DISPLACEMENT 5560 TONS.

	H.M.S. "Vanguard."			H.M.S. "Invincible."		
	Both Screws Full Power.	Both Screws Half Boilers.	One Screw Full Power.	Both Screws Full Power.	Both Screws Half Boilers.	One Screw Full Power.
Diameter of screw . . ft.	16·17	...	...	16·17	...	...
Pitch ,, . ,,	20·6 mean	...	...	17·2 mean	...	...
Revolutions per minute . .	73·7	60·27	70·49	79·0	62·18	79·0
Slip per cent. . . .	0·21 neg.	4·01 neg.	20·75 pos.	0·949 neg.	3·72 neg.	20·2 pos.
Speed of ship . . knots	14·944	12·742	11·356	13·51	10·926	10·797
Indicated horse-power . .	5366	2752	2903	4562	2438	2772
Indicated thrust . . lbs.	116,600	73,100	59,100	110,800	75,200	66,500
Tow-rope resistance of ship ,,	55,600	40,500	32200	45,620	29,800	29,140
Tr. horse-power . . .	2559	1583	1135	1891	1001	967
Efficiency, Tr. H.P. ÷ I.H.P.	0·479	0·575	0·392	0·415	0·410	0·349
Displacement $^{2/3}$ × speed^3 ÷ I.H.P.	195·3	236·0	158·4	166	169·5	139·6

Commodore Melville found by experiment on a twin-screw ship that with one screw running loose and disconnected from the engines for revolving it, when the ship was going 10 knots took 150 I.H.P.; whereas to revolve it and its engines coupled took 300 I.H.P.

Mr William Froude, when experimenting with the "Greyhound," found the resistance was less with the propeller secured than with it revolving—that is, the power to make it turn was greater than that required to drag it through the water masked behind the sternpost. Table XI. gives the results of some trials made by the Admiralty to test this question.

Two Screws.—Mr James Howden in 1874 fitted several large tug-boats with two screws, one at the bow and one at the stern, on the same line of shafting, operated together by one engine. They were said to be good at towing, and probably are handy in the sense that the bow propeller when running "astern" would soon check the headway of the ship, just as the floats of the paddle-wheel tugs do. This enables the tug to run close to its objective at high speed before slowing down. Further, with a tug, the back wash of the forward screw causing pressure on the bow and increase of skin friction is not of much moment *when towing*.

There is another arrangement of two screws which is noticeable and interesting, inasmuch as it is a survival of the invention of Perkins, Ericsson, and others. The Whitehead torpedo, in order that the movement of the screw may not affect the steering in the least degree, is propelled by a pair of screw propellers running co-axially in opposite directions, one before the other, on concentric shafts.

Triple screws were first tried on H.M.S. "Meteor" in 1855, with the view of getting a larger disc area and, consequently, more speed. There was, however, only one engine, which drove the centre screw direct, and it was geared to the shafts of the wing ones. The loss from friction and the faulty form of the ship were such that the speed was low and the efficiency very poor.

For the same reasons as given for preferring twin screws, especially with shallow-draught ships of high power, the arrangement of three screws naturally commends itself. It also has other advantages, however, inasmuch as by more subdivision smaller engines can be fitted, besides which when cruising there is a greater range of choice in the employment of the engines when comparatively low speeds are required. For example, the middle engine can be run at full power and with consequent high efficiency when the wing engines are standing; or the two wing engines only may be employed at full power, with a corresponding advantage.

With the advent of the turbine motor, however, the three screws became necessary in order that the propellers might be of the smallest diameter to keep their peripheral velocity within practical limits.

The Italian naval authorities were the first to give the triple screw system a practical trial by fitting it to the torpedo cruiser "Tripoli," 848 tons displacement, 2543 I.H.P. and 19 knots, built in 1886. They followed on with the "Montebello" in 1888, a sister ship with 19 knots speed, and later on some others.

Experiments had been made by the French naval authorities so far back as 1884-5, which resulted eventually in the building of the armoured cruiser "Dupuy de Lome" in 1890, by M. Marchal, chief constructor. This ship was very much larger than the Italian, being of 6400 tons displacement, 14,000 I.H.P., and having a speed of 20 knots.

In 1895 the French built a much larger ship with three screws, viz. the "Charlemagne," of 11,273 tons, 14,500 I.H.P. and 18 knots. Since then, most, if not all, of the French naval ships of large size have had the three-screw arrangement.

Commodore Melville, Engineer in Chief of the U.S. Navy, became a convert to and great advocate of the system, and in 1892 caused the cruiser "Columbia" to be so fitted. She was a comparatively large ship, being 412 feet long, 58·2 feet beam, 25·6 feet draught of water, 7375 tons displacement, 18,500 I.H.P., and on trial attained a speed of 22·8 knots. The Commodore studied the matter very fully, and was good enough to give his experience and the conclusions he had arrived at in a paper read to the Institution of Naval Architects in 1899. He advocated the making of the midship screw and machinery so that half the full power would be developed in them, and a quarter of the I.H.P. in each of the wing engines, and thereby to get a better variation in the division of power for cruising purposes. He estimated that the gain effected by the three-screw system over the twin screws was as much as 11·9 per cent. in the "Columbia," and that a fair average gain by using two screws instead of one screw would be 8 per cent. for speeds from 12 to 20 knots; that at 15 knots a ship will gain, by using three instead of two, 5 per cent.; while at 24 knots the gain in efficiency would be as much as 12 per cent.

These supposed gains were, however, called in question by the British naval authorities, and it was their opinion at that time that the practical objections to three screws outweighed any gain that was to be got, and until the turbine compelled them to depart from it, the twin screw remained the established practice in the British Navy.

In 1892 the German Government built the cruiser "Kaiserin Augusta," 6330 tons, 14,000 I.H.P. and 22·5 knots speed, and in 1897 some cruisers, somewhat smaller, all with three screws, followed by the "Fürst Bismarck," of 10,650 tons, 14,000 I.H.P. and 19 knots, an armoured ship, and since then several others.

Russia built the armoured cruiser "Rossia" in 1896, of 12,130 tons, 14,500 I.H.P. and 20 knots speed, and followed on with several other ships all having three screws. It may be said indeed that the Russians were the first to use three screws, seeing that the

Imperial yacht "Livadia," built by the Fairfield Company in 1886, had three screws.

Four screws have been fitted to the Cunard steamers "Lusitania" and "Mauritania" (see frontispiece), because of their great power and their high rate of revolution. It is not unlikely, from the success that has attended the arrangement in these vessels, that it may be followed by fitting in this way other and even smaller ships. By this subdivision reciprocating units of quite a reasonable size could be employed for very high power for two if not for all the screws, for there might be a reciprocating engine driving its own screw and exhausting to a low-pressure turbine, which in its turn would drive an independent screw on the other side, and *vice versa*, so that when cruising at half power or less a complete expansion installation could be used without affecting the steering of the ship.

The well-known Russian ice-breaker "Ermack" has four screws rather oddly distributed, viz. three at the stern and one at the bow. It is stated, too, that when the bow screw is going "astern" the greatest "ahead" effect is obtained—in fact, more than when it also was going ahead too. This curious result is attributed to the current set up around the bow and at the sides in the direction of the ship's motion by the fore screw when in stern gear.

Some of the Mersey ferry steamers have four screws, two at the bow and two at the stern, on a pair of shafts running through the ship on Howden's system. The vessels are double-ended and go either way. The "bow" screws are, as before mentioned, very effective in checking the headway on such ships, as well as in starting them quickly either way.

The position of the screw very materially affects its efficiency as a propeller. In the case of the "Archimedes" the propeller was a long one, and placed in a recess or gap in the deadwood (see fig. 4). It seems to have worked fairly well, considering all things, and the conclusion come to is that she had a clear "run"—that is, a very fine line after-body. The after-body of H.M.S. "Rattler" (see fig. 7) was all that could be desired for success as a screw steamer, and her high efficiency is no doubt in no small measure due to the position of the screw and the clean lead for the "feed" water. Whether this was not appreciated as it should have been by the naval constructors of the day, or whether parsimony overruled their science, it is difficult to say, but when the conversion of sailing ships into screw steamers was taken in hand some most egregious blunders were made, and although rectifying them was a costly business, it was an experience by which they and we benefit, or ought to. It was thought quite good enough to cut a gap in the deadwood immediately on the foreside of the stern-post, fit a new post, or in some way make a new end to the ship to which a stern tube was fitted and generally an arrangement made for a banjo frame, etc., so that the screw might be lifted up.

H.M.S. "Dauntless," a frigate of 2307 tons displacement, was in 1848 converted into a screw ship as described above, fitted with engines of 580 Nom.H.P. made by Messrs R. Napier & Sons, having two cylinders 84 inches diameter and 4 feet stroke, big enough to have driven her 11 knots, as the prismatic coefficient was only 0·72. As it was, on her official trial she only managed to do 7·366 knots with the engines developing 836 I.H.P. The tow-rope resistance was only 8378 lbs. and the Tr.H.P. 189·7. The efficiency, consequently, was only 0·227. A new stern was formed by carrying out the original lines under water so that the ship was 9·5 feet longer and the screw placed that amount of space further aft, which, although not much, was sufficient to effect a marked improvement in the speed, inasmuch as it was then 10·02 with 1388 I.H.P. and an efficiency of 0·350; with an alteration in the pitch of the screw, a better speed was maintained with 1217 I.H.P., and the efficiency rose to 0·432. (See Table XII.)

TABLE XII.—H.M.S. "DAUNTLESS" STEAM TRIALS BEFORE AND AFTER HAVING A NEW STERN, COMPARED WITH THOSE OF MODERN CARGO STEAMERS.

	Dauntless 1 as built 1848.	Dauntless 2 after Alteration. Same Screw.	Dauntless 3 after Alteration, with New Screw.	s.s. Z. Ordinary Cargo Steamer.	s.s. F. Ordinary Cargo Steamer.
Length beam draught .	210×40×16·3	219·5×40×16·6	219·5×40×16·3	250×30×16	240×33×15·0
Displacement prism coefficient	2307 - 0·723	2235 - 0·700	2307 - 0·708	2270 - 0·700	2350 - 0·750
Wetted skin	11934	12068	12136	12960	11334
Screw, diameter and pitch .	14·73×16·8	14·73×16·8	14·73×17·72	12·5×14·8	12·5×15·0
Revolution per minute	56·9	71·7	68·28	77·0	80·5
Slip per cent.	21·82	15·63	13·73	12·28	11·69
Speed . . . knots	7·366	10·016	10·293	10·00	10·55
Indicated horse-power .	836	1388	1217	685	866
Indicated thrust . . lbs.	27,588	38,170	33,050	19,760	23,890
Propeller thrust calculated	12860	18,420	17,680	13,575	..
Tr. resistance of ship . lbs.	8,378	15,688	16,629	12,960	12,581
Tr. H.P.	291	557	558	416	..
Tr H.P. . . .	189·7	483	525	398	407
Efficiency Tr. H.P. ÷ I.H.P.	0·227	0·349	0·431	0·586	0·470
Displacement$^{2/3}$×speed^3 ÷ I.H.P.	83·4	123·7	156·5	253·0	239

Two screws on the same shaft have been tried by Mr Parsons and others where limitation of draught or other causes have necessitated the use of a screw of such small diameter as to preclude the possibility of the necessary surface in one screw only for efficient propulsion. In such cases the screws are not close to one another as fitted by Mangin, but sufficiently far apart to have a feed for the aftermost independent of the stream from the leading one. The efficiency of such a combination must depend largely on the shape of the ship and the distance apart of the screws; hitherto the success of this arrangement of propellers has been somewhat qualified and not sufficiently pronounced to recommend its adoption in the future.

From Millington in 1816 and onward, inventors have not been wanting in taking out patents for so fitting the screw propeller that it shall help to, if not altogether, steer the ship; but very few of them have carried their ideas into practice. Fig. 45, however, is one of the few instances in which in workaday practice a second screw has been fitted abaft the propelling screw and arranged to help in steering as well as in propelling the ship. The illustration is taken from a photograph of the screw steamer "Stratheden," built in 1882, of 2000 tons and 200 nominal horse-power, and the arrangement of the auxiliary screw with its fittings and connections is in accordance with Mr J. J. Kunstadter's patent. It will be observed that the rudder pintles and gudgeons are of special design suitable for carrying the weight of the auxiliary propeller and its shaft, and the universal joint connecting the latter to the main screw shaft is of a substantial nature, as indeed it should be for the purpose. As the example set in this ship does not appear to have been followed, it is to be presumed that the advantages that the arrangement possessed were found to be outbalanced by the disadvantages, such as prime cost, exposure of auxiliary screw to damage from quay walls and entanglement with ropes and chains, as likewise the extra cost involved in examining and overhauling the propeller shafts, rudder, etc.

Scott Russell used to say that there were several places about a ship where a screw might be fitted, but that the worst place of all was the bow, because the stream delivered from the screw impinged directly on the bow and the current flowing about the bow and sides of the ship would be increased in velocity, causing a corresponding augment of resistance.

Professor Osborne Reynolds, and others before him and since his time, have taught and sometimes proved that the further away a screw is from the ship itself, the more efficient it becomes, and that when quite away from "wake" currents its maximum efficiency is reached. Possibly this teaching has induced the belief in the minds of some engineers that the bow is the place for a screw. This leads thought into another channel, and causes inquiry to be made into the nature, magnitude, and effect of the currents caused by the propeller itself on the ship.[1]

Propeller abaft the Rudder.—But long before Professor Reynolds made his researches, others had recognised the advantage of placing the screw as far away as possible from the body of the ship; for as early as 1855 steamers were built at Hull having the screw behind the rudder on the plan patented by John Beattie in 1850 (see fig. 44) and proposed by Ericsson in 1836.

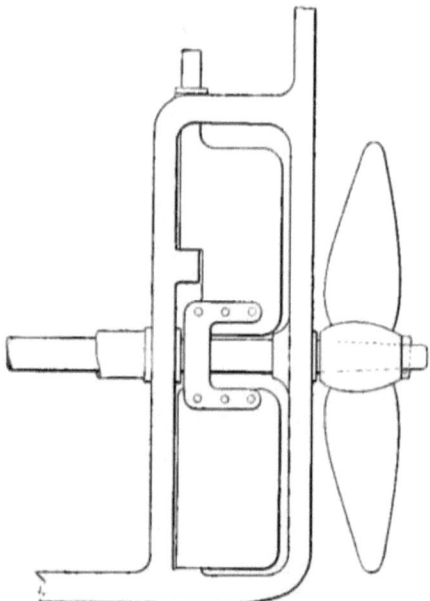

FIG. 44.—Screw outside of Rudder.

The "feed" to the screw so placed was excellent; and further, on reversal for "astern" motion, the impact of the stream from the screw was not direct on the body of the ship as is the case with the screw in its usual recess, when the result is that its motion is seriously checked; it is, of course, obvious that a screw standing out abaft the rudder-post must be exposed more than usual, and so liable to damage from hitting quay walls and pile work as well as to fouling with ropes and hawsers. It was on this latter account that fitting screws in this position was eventually discontinued and some of the ships so constructed were altered.

Submersible Screw.—To more than one of the early marine engineers the idea occurred of arranging the screw so that when the

---

[1] *Vide* Osborne Reynolds in *Trans. of Inst. Naval Architects*, vol. xvii., 1876; Geo. Calvert, *ibid.*, vol. xxviii., 1886, and vol. xxxiv., 1892.

THE NUMBER AND POSITION OF SCREWS. 149

ship was in open water with plenty of depth it could be more deeply submerged and its efficiency greatly improved thereby. It was then clear away from the broken wake of the bluff part of the after-body, and the possibility of the blades breaking through the surface was quite remote. Fig. 46 illustrates the method of affecting this change of position suggested by Shorter in 1800, by

FIG. 45.—Auxiliary Screw outside the Rudder.

Trevithick in 1815, Millington 1816, and patented by G. H. Phipps in 1850.

The late Sir Edward Harland carried the idea into practice on a big scale when he fitted the s.s. "Britannic," of Atlantic fame, with her huge propeller in a sliding frame, had the shaft connected to the next length by a massive universal joint so that when out of the Mersey and on the open sea it could be lowered down till the tips worked well below the line of keel. So much trouble was experienced, however, and when the risks of serious accident became appreciated, as they did very soon, that this magnificent ship was taken from her station and converted to the usual form of stern and propeller arrangement.

Submerged Screw.—Sir John Thornycroft and Mr Yarrow have for many years fitted torpedo boats and other small high-speed craft with the propeller so low down that its blades are far below the boat itself. This, however, was perhaps more in the nature, or perhaps the result, of a development, for, first of all, such fine-lined boats would have a long and deep "deadwood," producing a large amount of skin friction; it was gradually cut away till the shaft was reached, and, finally, the boat was designed with no deadwood and the shaft just inside the bottom. As these boats never have " to take the

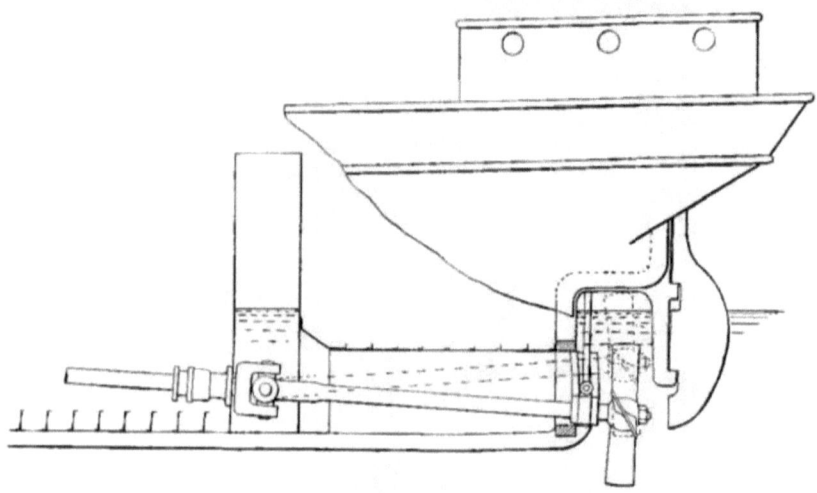

Fig. 46.—Phipps' Lowering Screw, 1850.

ground," and when on general service are in fairly deep water, there is no risk of damaging the screws, etc.

The position of twin screws was at first governed by the practice common with single ones: that is to say, they were placed just abreast of the spot where a single one would have been, that is, in a place just forward of the rudder so that that instrument worked quite free from the tips of the blades. The weight, etc., of the screws and their shafts were taken on A frames or brackets with two legs, one of which was vertical or nearly so, under the ship's counter, and the other horizontal (or nearly so) and attached to the ship's body. Sometimes the stern-frame had a gap, as for a single screw, so as to permit of the two propellers working with their tips nearly touching, and later on, in order to keep the screws as far in as possible so as to be protected by the ship's counters from damage, they were arranged to considerably overlap by placing one a few feet ahead of the other. The first twin-screw ship in the Navy, H.M.S. "Penelope," had double sterns united to one body and formed into one above water. That is, there were two runs each with deadwood and rudder as well as two screws. The ships so built were not at all satisfactory, and most costly to construct.

Of late years, especially with large ships, the after-body is so formed that no brackets are necessary. The two deadwoods or fins in this case are placed nearly horizontal and have at their outer edge the cylindrical formation enclosing and carrying the stern tube, shaft, etc. (See frontispiece.)

This, no doubt, is for all size of ships a much stronger design for carrying the twin screws, etc.; and, moreover, the shafts thereby are fully housed instead of exposed, and they can be more conveniently subdivided and handled. The flat formed by the two deadwoods or fins forms an admirable resistance to pitching, and so must tend to render the ship more comfortable in a head sea. On the other hand, a huge amount of surface is exposed to skin friction, and if the fins are not properly designed so as to follow the natural stream lines of the ship, they will form serious obstructions. To some extent also when pitching, even slightly, the feed to the propellers is affected by them.

The position of the triple screws differs somewhat from that of twin screws, inasmuch as the wing screws must considerably overlap the middle one, and as far as possible it is desirable that the latter should not injuriously affect the working of the wing screws or be affected by them. To this end the wing screws are well in advance of the centre one, which, of course, is only possible in ships of very fine lines and large beam. For this reason, no doubt, until the advent of the turbine with its small screw, the three-screw system was not followed in the mercantile marine.

It is pretty certain that the screws may with advantage be at a considerable distance forward, so long as the form and dimensions of the ship permit of it. In this case the method of carrying the wing screws, shafts, etc., will be the same as for twin screws.

Quadruple screws, as in the "Lusitania," are naturally in pairs (see frontispiece). The aftermost pair will be very nearly in the same plane as the single screw would have been in the case of triple screws, and the other pair will be forward of them, the exact position being determined partly by the disposition of the engines and partly by the shape of the ship. They may, in any case, be placed fairly close to the skin of the ship, and thereby work in the water that is influenced by the motion of the ship, which will there be free from eddies, and fairly constant in the direction and magnitude of its motion relatively to the ship.

CHAPTER XI.

SCREW PROPELLER BLADES: THEIR NUMBER, SHAPE, AND PROPORTIONS.

Number of Blades.—Various experiments have been made from time to time with model screws, with screws on steam launches and small yachts, and with screws on frigates and battleships, with the object of determining the relative merits of propellers with two or more blades, and, if possible, what number a propeller should have for highest efficiency. The late Mr Charles Sells, for so many years a most successful designer for Messrs Maudslay, Sons & Field, made an elaborate series of trials with screws having two, three, four and six blades varying in pitch, etc., the report of which is given in Chapter XVII.

Mr R. E. Froude has for many years devoted much time and thought to the study of the problems involved in screw propulsion, and made so many series of experiments in the tank at Haslar, that little remains for him to discover respecting either the form or number of blades of a propeller.

The late Mr Blechynden conducted a series of trials with model screws in a tank, or rather in an oval channel which permitted the water acted on by the screws to flow round and round without check or hindrance. The thrusts and turning moments were carefully measured and noted, and were published by him in a paper read to the members of the North-East Coast Institution of Engineers and Shipbuilders, 1886-7.

The following are the conclusions indicated by these careful and conscientious experiments:—

The thrusts on propellers of the same diameter, pitch, surface, and shape of each blade, and differing only *in the number* of blades, are as follows:—

Table XIII.

	Froude.	Blechynden.	Sells.	$\sqrt{\text{No of Blades}}$ 2
Screw with four blades	1·000	1·000	1·000	1·000
,, three ,,	0·850	0·862	0·907	0·866
,, two ,,	0·650	0·680	0·703	0·707

This really means that the thrust varies as the square root of the area of acting surface and not in direct proportion to it.

The following figures, taken from the trials of H.M.S. "Iris," seem to confirm this. In this case the first set of trials were made with four blades on each propeller; the second set were made after removing two opposite blades from each so that diameter, pitch, shape and area of each blade was the same in both trials, the aggregate area of acting surface being of course double in the first trial what it was in the second.

Table XIV.

Revolutions per minute.		40	50	60	70	80	90
Thrust with four-bladed screw	lbs.	20,000	31,000	45,000	60,000	78,500	98,500
,, two ,,	,,	14,000	21,500	30,000	41,000	55,000	70,000
Ratio of second to first .		0·700	0·69	0·67	0·69	0·70	0·71

Mr Isherwood had a series of trials made with a steam launch fitted with screws made by R. & W. Hawthorn over twenty-five years ago. Mr Blechynden made a careful analysis of them and came to the following conclusions as a result:—

(1) In screws of equal diameter and pitch, but of different blade area, *when the same thrust is developed, the turning moment is independent of blade area.*

(2) Screws of equal diameter and blade area, but varying in pitch ratios when tried under similar conditions and developing equal thrusts, have turning moments proportional to the pitch.

(3) Screws varying in diameter but of equal pitch ratios, developing equal thrusts, have turning moments proportional to the diameter.

(4) In any screw, if the total blade area remains constant and the blades are similarly shaped, the propelling effect is the same whether there are two or four blades.

This last conclusion is not quite in accordance with the deductions to be made from everyday practice, nor did Mr Walker,[1] who had made a series of trials with his yacht, agree to this proposition as being correct.

The third and fourth set of trials of the "Iris" may be appealed to, although perhaps objection may be taken to them, as they were a complete refutation of the above. But inasmuch as the two-bladed screws of the fourth trial with their surface of 56 square feet produced a better speed than the four-bladed ones of the third series with 72 square feet, it would seem to be contrary to what might have been expected with the fourth proposition before their eyes.

On the other hand, the trials of screws with different number of blades in H.M.S. "Emerald," while not confirming it, does not distinctly disprove it.

TABLE XV.—TRIALS OF H.M.S. "EMERALD" WITH VARIOUS SCREWS.

Particulars of Screw.		Common.	Common.	Common.
Diameter of screw	ft.	18·0	18·1	18·0
Pitch ,,	,,	28·0	26·0	26·0
Number of blades		Two	Four	Six
Developed surface		82·8	99·0	103·0
Revolutions		53·83	53·33	51·50
Slip per cent.		22·49	12·28	11·26
Speed of ship	knots	11·530	12·003	11·726
Indicated horse-power		2288	2323	2124
Indicated thrust		48,800	55,300	52,300

It is true that, here, area of surface is not exactly the same in all the cases, but the difference in area is not large, especially between the screws with four and six blades.

Mr Froude's experiments were carried out in the large tank at Haslar, with the model of the ship itself in advance of the model screw, and both moving together at the speed corresponding to the relative sizes of model and real ship. The thrust of the screw and the turning moments are most carefully taken by ingenious self-registering instruments, and every care taken to eliminate disturbing elements and keep the records absolutely correct.

Such experiments as carried out by these gentlemen are, of course, most interesting and likewise instructive, especially as enabling us to compare one model screw with another, but as a final criterion of the

[1] *Vide* Walker, *Trans. Inst. Naval Architects.* Also see Chap. XVIII.

screws most suitable for ocean-going ships there is good reason for reserve of judgment. It is by no means certain that an accurate comparison can be made between the performance of the actual screw in smooth water with that of its model in water of the same density and viscosity.[1] It is certain that in rough water with the ship subjected to a mixed motion of rolling and pitching even when slight, the performance of her screws cannot be gauged by model experiment in a tank. Such performance with the ship moving in any way but that of the straight line of its keel can scarcely be repeated in the tank, nor can the disturbing effects be measured or estimated. The comparative failure of the original screws of H.M.S. "Drake" and her sister ships is an indication of the fallibility of the tank method. At the same time all honour is due to Mr R. E. Froude and others for the very great—the almost incalculable—work they have all done both in general research and the practical service rendered for which they are primarily employed.

An experiment made on the s.s. "Charkieh" with a six-bladed screw of the same diameter, pitch, and surface as the three-bladed service one, and having the same shape of blade, is of interest, and gave the following results:—

Three-bladed screw. Coal consumption per day . . . 34·01 tons. Speed 10·69 knots.
Six-bladed screw. Coal consumption per day . . . 33·16 ,, ,, 10·65 ,,

Early practice with screw propellers was to follow the example set by F. P. Smith and Bennet Woodcroft, and fit two blades only instead of the multiplicity of Ericsson and others. This was no doubt largely, if not entirely, due to the fact that in H.M. service, as also in the mercantile marine, there existed the desire and need to retain sails as a means of assisting the machinery when at work and to navigate the ship when the wind served sufficiently well to do without the engines—perhaps also finally with the idea that the sails would always be a stand-by in a case of engine failures. Otherwise, there seems to have been no good reason for limiting propellers to two blades.

In the mercantile marine from the earliest days of steamships there had been a considerable number of them to which sails could be of little or no use, save, perhaps, only to steady them in a beam sea

---

[1] Froude's model screws are usually 8 to 9·6 inches in diameter.

and to prevent or quickly damp out heavy rolling. To such steamers a three-bladed screw would be an advantage, and no doubt such were used in comparatively early days, as well as the four-bladed ones were later on.

The blades of Smith and Woodcroft were for a considerable time about one-sixth to one-eighth of the pitch in length and the pitch fairly fine, consequently their tips were very broad, with the result that the greater part of the thrust was exerted near them. The difference between the pressure on the top and that on the bottom blade was therefore so great when at or near the vertical position that the vibration was very considerable in amount and trying in character. This was, of course, aggravated when the screws were of large diameter and only just immersed. Screws with two blades by Griffiths, Sutherland, and others, having comparatively narrow tips, did not so much distress either ship or passengers. Screws with more than two blades, even when fairly broad tipped, were not so bad as the two-bladed variety.

The Admiralty, be it recorded to their credit, made some interesting and valuable experiments to assure themselves on these points, chiefly on the best number of blades for naval purposes (*vide* Chapter XVI.). In 1861 the three-decked battleship "Duncan," of 3985 tons, was tried with a three-bladed screw in place of other two-bladed ones; later this screw had its leading corners cut off as to reduce the area of its acting surface from 115 square feet to 107·8, as the two-bladed Griffiths had only 77·6 square feet. The speeds varied very slightly, the highest being with the Griffiths, but the speed coefficient with it was only 176·2 as against 187·6 with the three-bladed.

A year later further experiments were made with screws differing chiefly in number of blades on H.M.S. "Shannon," a frigate of 3612 tons. In this case the screws having more than two blades were on Woodcroft's principle and had blades of the same size, diameter, and pitch, so that the acting surface of the six-bladed screw was double that of the three-bladed one. The two-bladed one was, however, a common one, but with surface nearly proportional to the number of blades, as were the others.

The best speed, 11·55 knots, was made with the four-bladed screw, the three-bladed running it close with 11·492 knots. The efficiency as measured by the speed coefficient gives the same verdict in favour of the four-bladed screw.

In 1863 a further experiment was made with H.M.S. "Emerald," a frigate of 3563 tons (see p. 155). In this case the propellers were two-, four-, and six-bladed common screws with surface 82·8, 99·0, and 103 square feet, a much more satisfactory proportion. The highest speed, 12·003 knots, was attained with the four-bladed, with a speed coefficient of 171·7; the next highest being 11·726 knots, with a speed coefficient, however, of 175·1 by the six-bladed screw. Following these trials a considerable number of ships were fitted with four-bladed screws, but with the results already stated.

The modern practice as to number of blades on a screw may be summed up by saying that in the mercantile marine the four-bladed screw is almost universally employed; in the naval service and with all modern swift express steamers, three blades are almost always employed. In both services, however, are still to be found propellers with two blades notwithstanding the prevalence of these rules. Now, as a matter of fact, the right number of blades for a particular propeller cannot be determined by fashion nor be a matter of prejudice, but must be governed by the conditions impressed on and ruling it only.

Two-bladed screws have some special claims for consideration. In the experiments with H.M.S. "Duncan," "Shannon," and "Emerald," the performances of the two-bladed screws were by no means bad by comparison with the others. An inspection of the results of these trials impresses one with the idea that had the two-bladed screws been run at the same revolutions and their surface been nearly equal to that of the other screws, very different verdicts would have followed. Even as it was, with H.M.S. "Duncan," the better designed Griffiths with its two blades gave the best speed. Its low-speed coefficient is, doubtless, due to the high speed of revolution. Griffiths' improved screws with two blades were very efficient and the vibration quite moderate, even if due to them.

Now, as a matter of fact as well as experience, the two-bladed screw does not require so much total acting surface as one with more blades, because of the greater efficiency of blades, due to the breadth; besides which, there are fewer resisting edges. If of the same surface the breadth of blades will be as 50 to 33 of a three-bladed screw. A reduction of 10 per cent. in surface with the two-bladed screw means a substantial saving in surface friction, while the blade breadth is still as great as 45 to 33, and thus possessing a superiority sufficient to produce a marked difference in results.

It is, therefore, no matter for wonderment that with its superiority in efficiency the Griffiths screw was retained by the Admiralty for all classes of ship. In spite of the claims of other propellers, to the present day, so far as shape is concerned, the Griffiths idea is followed. In the case of the "Iris" the performance of the original screws with two of the blades of each removed was very good, compared not only with that of the screw bearing the whole four blades, but even when judged by that of the new propellers. In fact, had those blades been given a foot or two more pitch there would have been no justification for new ones, except, perhaps, on the ground of vibration. Again, it is noticeable that of the new screws made specially to replace the condemned ones, the two-bladed Griffiths, in spite of its huge diameter, gave the highest speed with the best speed coefficients.

To-day we have yachts and other cruisers for waters where coaling stations are few and far between, and other ships with which quickness of passage is not so important as cheapness. Such ships continue to be fully masted and have full sail power, and yet require screws which combine minimum of obstruction with the maximum of convenience for adjustment so as to yield the best combined steam and sail results. They are therefore generally fitted with Bevis' feathering screw (see fig. 47) which allows of this, inasmuch as the two blades can be turned so as to partly fill the stern apertures and yet be masked by the stern-post.

The motor boat, which has come to stay, in spite of its non-reversible engine and many noises and smells, requires a somewhat similar screw, inasmuch as, since the engine can only run one way, the propeller must do the same, unless the objectionable wheel gearing is introduced; the screw blades must then turn round so as to become a left-handed screw if in "ahead gear" it was right-handed. Such screws will of necessity be two-bladed and have a flat acting surface on the aft side; but probably slightly convex surfaces on both sides will be the most efficient under such circumstances.

Maudslay's Feathering Screw and Banjo Frame.—In the early days of the screw frigate it was desirable to have a propeller whose pitch could be altered, and to have an apparatus for lifting it out of the water, so that it ceased to obstruct the passage of the ship. Most of these old ships sailed remarkably well, and in order to use the steam power to advantage when the ship was doing well under sail,

160   MARINE PROPELLERS.

it was necessary to increase the pitch of the screw so as to keep down the revolutions of the engine. Several inventors took out patents for the purpose of attaining this end, and one of the best and

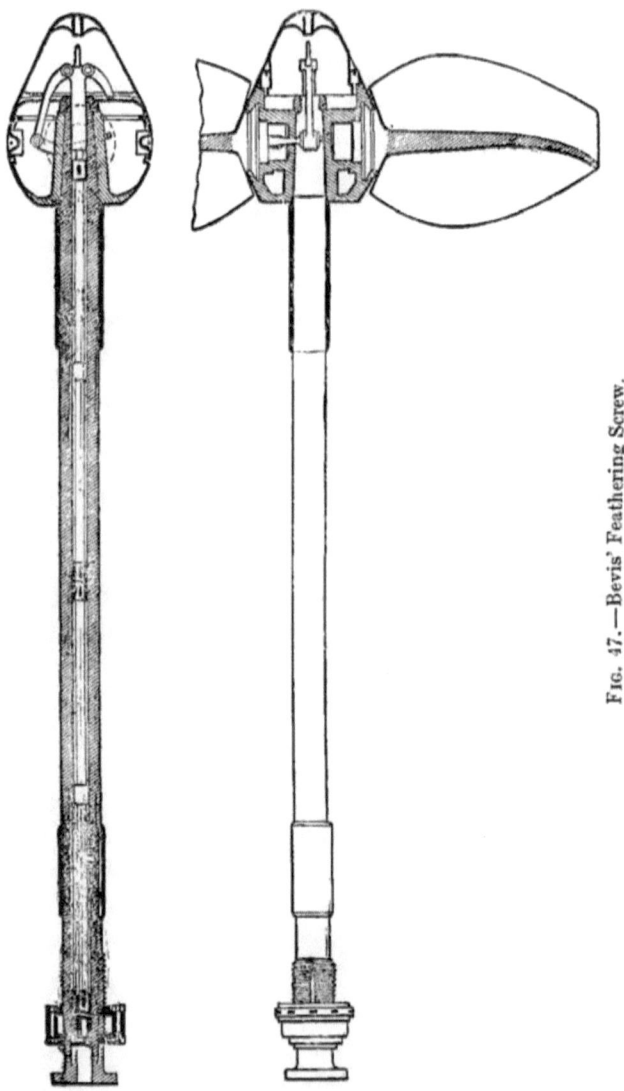

Fig. 47.—Bevis' Feathering Screw.

most successful was that fitted by Messrs Maudslay, Sons & Field to H.M.S. "Aurora" and other ships. Fig. 48 shows very clearly the method of accomplishing the end in view, and also the banjo frame containing the screw, and in which it was raised to deck level when the ship was under sail and lowered into position and secured there when under steam.

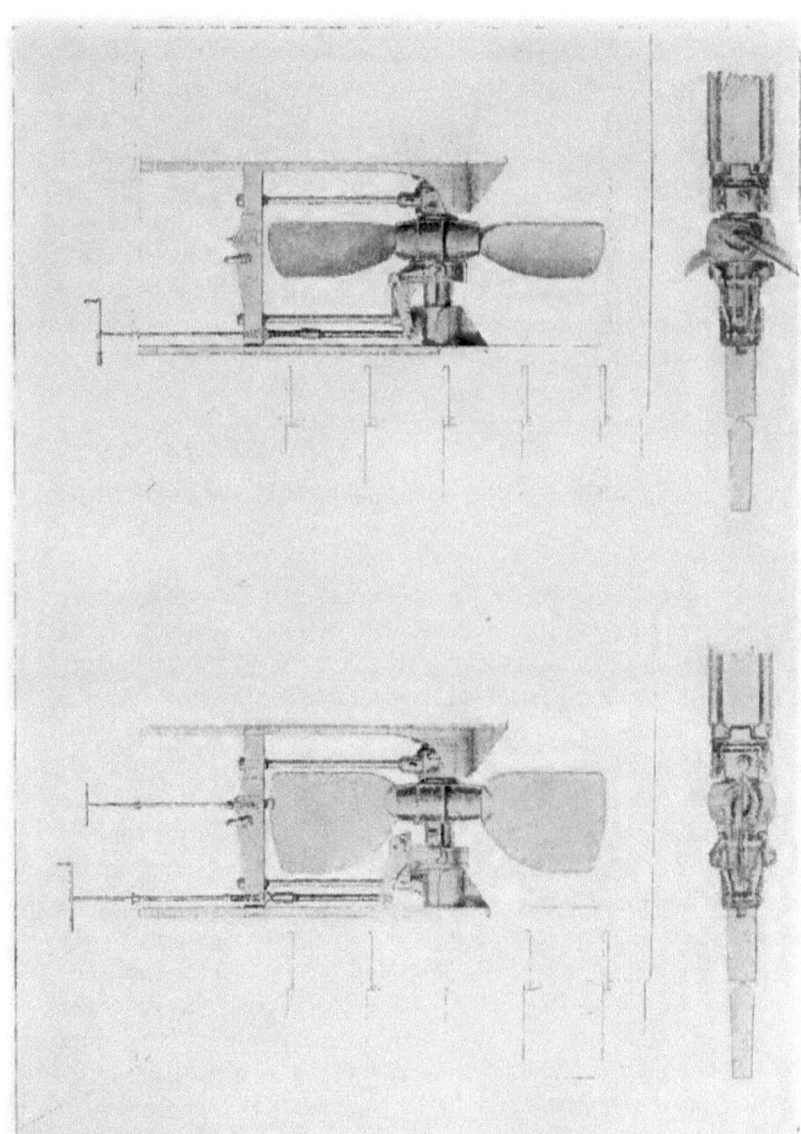

Position when sailing.   When Steaming
Fig. 48.—Maudslay's Feathering Screw and Banjo Frame.

The method adopted by Mr Bevis of Birkenhead, which is shown in fig. 47, was, however, so much neater and easier to deal with, that in course of a few years it remained as the only feathering propeller in use, as it still is to-day. It will be seen that both Maudslay's and Bevis' screws could have the blades turned to the fore and aft position so as to cause no obstruction when sailing.

Flat blades, that is, those having a constant angle and consequently a pitch increasing from root to tip, have been tried on propellers where complete feathering, that is, the blades of which are turned exactly fore and aft, is required for efficient sailing; or, in cases where the reversing cannot be done by the motor, but must be done by reversing the screw from a right-handed to a left handed-one, such blades would seem to be desirable. But, as a matter of fact, a propeller fitted with them is of very low efficiency, and is equally bad in either head or stern gear; while a helical blade may lack efficiency when reversed for stern gear, at which times it is not important, as they are of short duration and not frequent. Sir John Thornycroft found that such a screw required nearly twice as much power as a helical screw of the same dimensions. At $11\frac{1}{2}$ knots the flat blades required 173 I.H.P. to drive them, whereas 90 I.H.P. was sufficient for the common screw; and at 10 knots there was practically the same difference.

Three-bladed screws are, on the whole, the most satisfactory for general purposes, for they possess high efficiency when working under almost any circumstances; that efficiency is satisfactory whether the screw be working at a considerable depth or so near the surface as to induce air currents to follow it. The increase of total blade surface over that of the two-bladed screw is not, after all, necessarily very large, while its breadth of blade is respectable, being, as already stated, 33 as against 45, or a reduction of 26·6 per cent.

In the matter of vibration the three-bladed screw is superior to either the two or four-bladed variety, inasmuch as no two of its blades are opposite one another so as to form a couple tending to shake the stern, as is the case with both the two- and four-bladed screws.

The loss of a blade amounts of course to 50 per cent. reduction with a two-bladed as against 33 per cent. with the screw having three blades; but as the two blades of the one screw have a smaller surface than the three of the others (say by 10 per cent.) the comparison is 33 against 45 per cent.

Four-bladed screws have of late years found very little favour either in the Navy or with the designers of high-speed steamers of all sizes. Even in the case of the huge Cunard steamers "Lusitania" and "Mauritania," with their great power and small diameter of propeller, three blades only were at first fitted. After several voyages, however, it was found advisable to reduce the pressure per square foot on the blade surface, and for this and other reasons new screws having each four blades have been fitted, and proved advantageous. With practical sea-going engineers of the mercantile marine, young and old, it is an axiom that a sea-going, that is, ocean-crossing, ship, must have four-bladed screws, and that monster with its four great wings has almost become a fetish to many of them.

They claim, and rightly, that the loss of a blade is a shortage of 25 per cent. only as against 33 of the three-bladed screw, but, on the other hand, with twin-screw ships and three-bladed propellers the loss of a blade means a shortage of only 16·5 per cent. of total effective surface. A modification of the percentage of loss is requisite here also, from the fact that in practice the surface of the four-bladed screw is usually larger than that of the three-bladed, all other things being the same. It is also necessary to bear in mind that with the larger surface there is the greater loss from friction; and further, that with the increase in number of blades there is the inevitable increase in the resistance to rotation. It will be seen, then, that for the same reason that the two-bladed is superior to the three, the four-bladed is inferior to that screw. But in rough or moderately rough weather the four-bladed screw is the better for single-screw ships, even when of comparatively small sizes. For twin, triple and quadruple, in which, owing to the screws being of smaller diameter and their centres lower, the propellers are better immersed, the three-bladed screw is the best for all sorts and conditions of service.

In small ships, especially those employed in sheltered waters, where the waves are only of small dimensions, even two-bladed screws of the Griffiths type may be used with advantage. In very small vessels two-bladed screws may always be used.

The shape of screw blades best suited for each type of ship and condition of work is a problem that may be worked out some day by some one whose leisure equals his knowledge of mathematical science. Judging by everyday experience, however, there does not appear to be much scope for anything but small refinements, and it is very doubtful if such limited differences seriously affect the efficiency of a screw for any other reason than a quantitative one of acting surface.

It may be taken as an axiom that no screw is satisfactory that works with excessive vibration, that is, which itself produces vibration in the hull of a ship distinct from the vibration set up by the unbalanced moving parts of the engine itself.

It was found in old days that while a screw with broad tips, such as the "common screw," could drive the ship at good speed and be otherwise efficient, the vibration was always high compared with that of the narrow-tipped screws of Griffiths.

Hence it is another axiom that the propeller, to be successful, must have either a rounded end or a narrow tip to the blades.

These axioms apply more forcibly to the propellers of excessive diameter so often found in ships of all kinds up to twenty years ago than to the modern ones, whose diameter is generally erring on the side of smallness. It is therefore now the practice with such screws to broaden the tips very considerably, but never to make them square.

The maximum breadth was usually at a distance of $\frac{1}{4}$ to $\frac{1}{3}$ of the diameter from the axis; the breadth at the boss was somewhat less, generally about $\frac{8}{10}$ths of the maximum, the length of the boss being such as to take this. The breadth of the tip would be about five to six tenths of the maximum. A curve then would be drawn through the terminals of these breadths, that portion between the boss and maximum having a much larger radius of curvature than the portion from the maximum to tip.

Griffiths' rules for screw blades are very precise and were closely followed for a great number of years; and bearing in mind that they were and are prescribed for two-bladed propellers, they do not in essence differ from the general practice of to-day. They are (*vide* fig. 38) as follows:—

Diameter of the boss to be from $\frac{1}{3}$ to $\frac{1}{4}$ the diameter of screw.
Diameter of the flange of blade to be $\frac{1}{2}$ the diameter of boss.
Thickness of the flange of blade to be $\frac{1}{20}$ the diameter of boss.
Width of blade at widest part to be $\frac{1}{3}$ the diameter of screw.
Width of blade at tip to be $\frac{1}{4}$ the diameter of screw.
Thickness of (bronze) blade at root to be $\frac{1}{36}$ the diameter of screw.
Thickness of (bronze) blade at tip to be $\frac{1}{8}$ the thickness at root.
Diameter of shank to be $\frac{1}{4}$ the diameter of screw.
Curvature of blade forward to be $\frac{1}{24}$ the diameter at the tip.

The form of blades for a propeller having three on modern lines will be found by the following rules:—Sf is the surface ratio; the diameter of boss must not be less than three times the diameter of shaft.

Maximum breadth of blade at a distance of 0·25D from the centre $=\dfrac{Sf}{1\cdot 35}$ of the diameter of screw.

Breadth at tip $= 0\cdot45 \times$ maximum breadth.

If the propeller has four blades,

$$\text{Maximum breadth} = \dfrac{Sf}{1\cdot 62} \text{ of the diameter of screw.}$$

The acting surface of a screw propeller may be found by the formula already given for determining the thrust of a screw, viz.,

$$\text{Thrust in pounds} = \dfrac{D \times \sqrt{A} \times V^2}{P_r} \times G,$$

where D is the diameter in feet, A the total area of the acting surface of all the blades in square feet, V the velocity of the screws in feet per second, P_r the pitch ratio, and G a factor depending on shape of the blade. (*Vide* page 124.)

Rule 1.—$A = \left(\dfrac{T \times P_r}{D \times V^2 \times G}\right)^2.$

As some of the above figures are not always available, and as the amount of surface is modified somewhat by the form of the ship, and also the position of the screw affects its efficiency, the following empiric rules may be followed in calculating the surface, etc., of screw propellers so as to give satisfactory results agreeing with good practice.

Rule 2.—Area of acting surface $= K\sqrt{\dfrac{I.H.P.}{\text{Revolution}}}.$

$K = M \times$ prismatic coefficient of fineness of ship.

The prismatic coefficient must not be taken at less than 0·55.

For four-bladed screws, single screw, M is 20, and for twin screws 15
,, three- ,, ,, ,, 19, ,, ,, 14
,, two- ,, ,, ,, 17·5, ,, ,, 13

Example 1.—A twin-screw steamer whose speed is 20 knots; her engines develop a total of 6000 I.H.P. at 150 revolutions to get this speed; her prismatic coefficient of fineness is 0·600, and her screws are to have three blades.

In this case $K = 14 \times 0\cdot 6 = 8\cdot 4.$

Area of each screw $= 8\cdot4\sqrt{\dfrac{3000}{150}} = 37\cdot 55$ square feet.

Example 2.—A single-screw steamer whose prismatic coefficient of fineness is 0·72 is steaming at 15 knots with engines running at 75 revolutions and indicating 4500 H.P. Her propeller has four blades.

In this case $K = 20 \times 0·72 = 14·4$.

Area of screw blades $= 14·4 \sqrt{\dfrac{4500}{75}} = 111·5$ square feet.

Example 3.—A turbine steamer having three screws is to steam at 25 knots with turbines running at 500 revolutions per minute and developing a power equal to 12,000 I.H.P. Her prismatic coefficient is 0·56.

Here $K = 14 \times 0·56 = 7·84$.

Area of surface of each side screw $= 7·84 \sqrt{\dfrac{4000}{500}} = 22·2$ square feet.

Example 4.—A steam yacht whose speed is 12·5 knots is to be driven by a two-bladed screw making 250 revolutions per minute and indicating 750 I.H.P. Her prismatic coefficient is 0·58.

Here $K = 17·5 \times 0·58 = 10·15$.

Area of screw blades $= 10·15 \sqrt{\dfrac{750}{250}} = 17·55$ square feet.

In these rules and examples it is taken for granted that the diameter will not be abnormal, that is, that it will have been deduced from the rules laid down herein for diameter (see p. 171).

On the foregoing basis the following holds good :—

Rule.—Maximum breadth of blade in inches $= N \sqrt[3]{\dfrac{\text{I.H.P.}}{\text{Revolutions}}}$.

The value of N for a four-bladed screw is 14, for a three-bladed 17, for a two-bladed 22.

(*a*) The maximum breadth of blade in Example 2 will be

Maximum breadth $= 14 \sqrt[3]{\dfrac{4500}{75}} = 54·7$ inches.

(*b*) That in Example 4 :

Maximum breadth $= 22 \sqrt[3]{\dfrac{750}{250}} = 31·6$ inches.

The diameter of boss was in the early screws quite small ; in fact, a mere cylinder of metal of sufficient thickness to resist the tear of the blades. Griffiths was the first to recognise the futility of such bosses and to perceive that the blade at its root was so nearly fore and aft that its propelling power was quite small compared with its resistance to rotation. Sunderland and others had recognised the loss from this cause, and proposed to remedy it by cutting away the blade till it became a mere stem or root of the acting surface, which was in their case quite remote from the boss. Griffiths removed the superfluous and harmful portion of the blades by enclosing it in a huge spherical boss which displaced the water, which otherwise was merely churned about and made to produce eddy currents. The Admiralty tested the effect of a large boss by an interesting trial with H.M.S. "Conflict" of 1750 tons in 1853.

TABLE XVI.—TRIALS OF H.M.S. "CONFLICT" WITH DIFFERENT BOSSES.

Number of Trial.		No. 5. Plain Boss.	No. 6. Large Boss.	No. 7. Plain Boss.
Speed	knots	8·837	9·425	9·424
Revolutions	.	73·0	77·0	75·75
I.H.P.		752	784	812
Slip	per cent.	19·08	18·18	16·84
$S^3 D^{2/3} \div$ I.H.P.	. .	133·4	154·5	149·8

A large spherical wooden boss was fitted to the boss of the common screw with the leading corner of its blades cut away. The 6th and 7th trials were run in smooth water on a calm day, and the 5th in slightly windy weather. It will be seen by these figures that there was a distinct gain with the enlarged boss. This test, together with the adoption of loose blades, whereby a large boss is necessary, caused the Admiralty to adopt the large sphere, while the mercantile marine with its solid cast-iron propellers stuck to the comparatively small egg-shaped boss, and, as a matter of fact, practically do so now, for it is only in the case of loose blades that a large boss is found on a merchant ship.

The diameter of the boss with loose blades depends very much on the number of blades; its size therefore should be determined by the following formula :—

Diameter of boss $= 0.5 \sqrt{\text{diameter of screw} \times \text{number of blades}}$.

Example.—To find the diameter of boss suitable for a loose-bladed propeller 16 feet diameter, (i.) with two blades, (ii.) three blades, (iii.) four blades.

(i.) diam. $= 0{\cdot}5 \sqrt{16 \times 2} = 2{\cdot}83$ feet.
(ii.) ,, $= 0{\cdot}5 \sqrt{16 \times 3} = 3{\cdot}46$,,
(iii.) ,, $= 0{\cdot}5 \sqrt{16 \times 4} = 4{\cdot}00$,,

For the small diameter screws with high power now common in the naval and mercantile express service the multiplier should be 0·7 instead of 0·5, so that had the example been for such ships the diameters would be 4·0, 4·85, and 5·6 feet respectively.

This means that the diameter of boss with loose blades should be not less than 2·7 × diameter of screw shaft for three blades and 3·2 × diameter of screw shaft for four blades.

The length of boss should not be less than twice the diameter of the screw shaft, and with a solid shaft it should be 2·5 × diameter of shaft.

The diameter of boss of a solid screw should be not less than 2·25 times the diameter of shaft; nor the length than 0·2 × diameter of the screw, exclusive of any nutshield in rear or gland cover in front. It should then be 2·2 to 2·5 × diameter of shaft.

Elongated Bosses.—It has been pointed out that so early as 1851 one Roberts suggested putting a conical tail to the boss so that there should be no loss there from eddies, and to prolong the boss forward so as to come in line with the ship's form and so avoid loss. It was very many years after that Thornycroft adopted this method for increasing the efficiency of the screw.

CHAPTER XII.

DETAILS OF SCREW PROPELLERS AND THEIR DIMENSIONS.

The Diameter of the Screw Propeller is one of the three features by which its efficiency is governed, the others being pitch, acting surface and shape of blade, or rather the disposition of the surface. Consequently it is easy to understand that there may be a considerable variation in the value of each, without much disturbance of efficiency in any screw running free from the influence of external bodies such as the ship herself, or the sea bottom or river banks near which it operates.

For particular cases, such as a tug-boat, where thrust *per se* is of first consideration and therefore large diameter necessary, and in the case of turbine-driven screws where large diameter is impossible on account of excessive peripheral velocity, this feature of a propeller must be the leading one; in general cases there may be almost any reasonable diameter of screw without violent variation in efficiency, provided the other features are modified to suit as in the formula:—

$$\text{Thrust in pounds} = \frac{\text{diameter} \times \sqrt{\text{area of surface}} \times v^2 \times f}{\text{pitch ratio}},$$

and the thrust v and f are assumed to be constants.

In the past the disposition, due largely to Rankine's teaching, has been to fit a screw of as large a diameter as the ship's arrangements permit; to-day the tendency is quite the other way, and it is now the practice of the expert engineer to try and find which of the many sizes is most suitable for his particular ship on the ground of giving the highest speed to the ship, or of propelling the ship with the maximum efficiency, which is not always the same thing, as may be seen by referring to the various trials whose records are given in Chapters XVII. and XVIII.

The following, which are practically axioms, should be considered carefully in making a decision :—

1. Theoretically a screw of large diameter is most efficient.
2. A large screw, especially if of fine pitch, tends to make serious augmented resistance, which means loss of efficiency.
3. It generally runs with *negative* apparent slip.
4. The skin and edge resistance of the blades is large.
5. The tip of the upper blade is so near the surface as to cause breakage of water on slight disturbance, with consequent aeration of " feed " and loss of efficiency.
6. Liability to hit wharfs and quay walls, resulting in damage to blade tips.
7. A small screw well immersed has a higher efficiency than a larger one near the surface, so much so that even its propelling effect is higher.
8. There is not necessarily any specific relation between the disc area of screw and immersed midship section of the ship; each may vary in its own way.
9. By special formation of the stern in the neighbourhood of the screw, the diameter may be considerably in excess of the draught of water.
10. The screw propeller is influenced as to efficiency by the position on the ship and the form of the ship.
11. The diameter must be modified to suit each ship.
12. The diameter of a twin screw may be less than that of a single screw, all other things being equal; that is, when driven by an engine of the same size, with the same boiler pressure, and being of the same pitch as a single screw, each twin may be of less diameter and correspondingly less pitch. In the case of quadruple screws, for similar reasons, their diameter may be modified considerably.

Taking all these things into account it would seem that the propeller must be chiefly governed by the *torque*, that is, the turning power exerted by the engine, and be modified by the circumstances surrounding it. It was customary at one time to arrive at the proper diameter of screw by taking the ratio between it and the length of stroke of piston, which ratio was and is now about 3·5 in fast steamers and 4·0 in cargo ships. Again, there is, as a matter of fact, a nearly constant ratio between the diameters of the screw and low-pressure cylinder, viz. 3·0 to 3·25, but the following rule may be taken, as giving a diameter of screw appropriate to the size of the engine and modified to suit the ship.

DETAILS OF SCREW PROPELLERS.

D is the diameter of low-pressure cylinder in feet; when there are two L.P. cylinders of diameter D, then $D = 1.4 \times D$.

S is the stroke of piston in feet.

Pc is the prismatic coefficient of the ship.

Z is a multiplier $= (2.4 + Pc)$ for twin screws, and $(2.7 + Pc)$ for singles.

RULE 1.—Diameter of screw in feet $= Z \times \sqrt{D \times S}$.

(a) *Example.*—To find the diameter of screw of a yacht whose prismatic coefficient is 0·67 and engines with cylinders 18 and 36 inches diameter and 24 inches stroke.

$$\text{Diameter} = (2.7 + 0.67)\sqrt{3 \times 2} = 8.25 \text{ feet.}$$

(b) *Example.*—What size propeller should an express twin-screw steamer have whose coefficient is 0·54 and her engines with cylinders 30, 45, 51 and 51 inches diameter and 36 inches stroke?

$$\text{Diameter} = (2.4 + 0.54)\sqrt{6 \times 3.0} = 12.47 \text{ feet.}$$

(c) *Example.*—How large a screw should a tramp steamer have whose cylinders are 25 to 39 and 64 inches diameter and 45 inches stroke, her prismatic coefficient being 0·8?

$$\text{Diameter} = (2.7 + 0.8)\sqrt{5.33 \times 3.75} = 15.65 \text{ feet.}$$

(d) *Example.*—The diameter of the screws of a twin-screw cruiser whose coefficient is 0·57 and has engines having cylinders 43, 69, 77 and 77 inches diameter and 42 inches stroke.

$$\text{Diameter} = (2.4 + 0.57)\sqrt{1.4 \times 6.4 \times 3.5} = 16.54 \text{ feet.}$$

The size of engine is not always known in the early stages of the design of a ship, hence another method of getting an appropriate diameter is desirable; moreover, the above rule is not applicable to turbine-driven screws.

The following rule will give such appropriate or suitable size of screw that it may be followed in designing both ship and propeller.

RULE 2.—Diameter of screw in feet $= x \times Pc \sqrt{\dfrac{I.H.P.}{R}}$.

In this case the following are the values of x:—

For single screws $x = 7.25$, and for ocean-going expresses 7·61
" twin " $x = 6.55$ " " " " 6·88
" quadruple " $x = 6.25$ " " " " 6·51
Turbine-driven centre screws $x = 6.55$ " " " " 6·88
Turbine-driven wing screws $x = 5.75$ " " " " 6·04

With this rule Pc must not be taken at a less value than 0·55, and for ocean-going ships whose normal speed is high and approximating to that of full power, the value of x may be increased by 5 per cent. as shown above.

Example (a).—The diameter of the screw of a torpedo boat whose prismatic coefficient is 0·55 and I.H.P. 2000 at 400 revolutions per minute =

$$7\cdot 25 \times \cdot 55 \sqrt[3]{\frac{2000}{400}} = 6\cdot 8 \text{ feet.}$$

(b.)—What is an appropriate diameter for a cross-channel twin-screw express steamer of coefficient 0·60, each engine indicating 3000 I.H.P. at 150 revolutions?

$$\text{Diameter} = 6\cdot 55 \times 0\cdot 6 \sqrt[3]{\frac{3000}{150}} = 10\cdot 65 \text{ feet.}$$

(c.)—A triple screw Atlantic steamer whose prismatic coefficient is 0·65 has engines of 3000 I.H.P. divided equally between the screws, which run at 100 revolutions. The appropriate diameter of the centre is $7\cdot 61 \times \cdot 65 \sqrt[3]{\frac{10,000}{100}} = 23$ feet, and the diameter of each wing screw =

$$6\cdot 88 \times \cdot 65 \sqrt[3]{\frac{10,000}{100}} = 20\cdot 7.$$

(d.)—What should be the diameter of the wing screws of a quadruple screw turbine-driven Atlantic steamer, each one having 16,000 I.H.P. at 180 revolutions delivered to it? Prismatic coefficient 0·64.

$$\text{Diameter} = 6\cdot 04 \times 0\cdot 64 \sqrt[3]{\frac{16,000}{180}} = 17\cdot 26 \text{ feet.}$$

Having fixed on a diameter, the general design of a screw can be gone on with after the area and pitch have been ascertained as appropriate to and suitable for it with the conditions under which it has to work by the rules already given.

The screw blade is subject to stresses at its root, due to the thrust on it which produces a bending moment and a shearing force. It is also subject to bending and shearing due to the resistance to rotation; and, further, has stresses due to the centrifugal forces. Not only is the blade root subject to such stresses, but at every section outwards, till the tip is nearly reached, they exist.

It is usual, and very properly so, to teach in colleges and technical schools how these stresses may be calculated at each section, or rather how the forces acting on each section may be calculated with a more or less approximation to accuracy, on the supposition that the screw is working at uniform velocity in smooth water; and in this way step by step to determine the thickness of a blade throughout for good and safe working under such conditions.

But what the nature of the forces is or what the magnitude of the stresses borne daily by propeller blades in an Atlantic steamer amount to, is beyond the power of the mathematician to estimate, and will probably always remain unknown, in spite of the fact that the propeller of such ships must always be designed and made strong enough to withstand them. Further, if it were possible, it would take too long a time to calculate the stresses at, say, six different sections of a propeller blade in order to decide on their thickness; besides which, even when done, it is more than possible that the error factor of the process could be quite as large as that involved in making the calculations by empiric formulæ.

The thickness of propeller blades may, however, be determined by calculations with a fair degree of accuracy by assuming that the *maximum load* on them is one and a third times the mean thrust, as calculated by the rule given on p. 124, or the indicated thrust may be taken as the gross, when it can be obtained, as the measure of the maximum load. It must also be assumed that this load is borne uniformly by and over the acting surface of the screw. Then if As is the acting surface of the screw in square inches and I.T. is the indicated thrust in pounds,

$$\text{Max. pressure } p \text{ on the screw blade} = \text{I.T.} \div As.$$

To find the thickness of a blade at any points from the tip to the root it will be sufficient to calculate the bending moment on section at B, C, and D, fig. 49, and at corresponding points XYZ on a radial line ZK, to draw ordinates XN, YM, and ZQ, which in length represent, on a convenient scale, the magnitude of the bending moment at those points. That at any other point is measured by the length of ordinate intercepted by the line OK and the curve QNK.

To find the bending moment on any section, say B, it will be necessary to imagine the blade surface from B to A divided into a number of narrow strips—say an inch wide. Multiply the length of one EF ($=b$) in inches by the pressure p, when the load will be $b \times p$. This is acting at a distance BG ($=x$) so that the bending moment $= (b \times p) \times x$ inch pounds.

The B.M. due to each strip can be calculated in the same way, and finally, when all are added together, the sum is the bending moment due to the pressure on the surface BEAF.

Now if b is the breadth of blade at a section of which it is desired to know the thickness t and B.M. is the

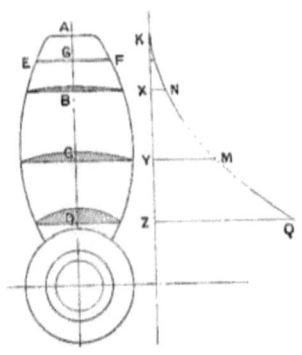

Fig. 49.—Showing Curve of Bending Moments.

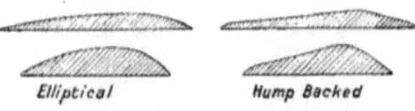

Elliptical   Hump Backed

Fig. 50.—Typical Screw Blade Sections.

bending maximum moment on the section, then

$$t = \sqrt{\frac{\text{B.M.} \times f}{b \times s}} \qquad . \qquad (1)$$

f is a factor which, for semi-elliptical sections, may be taken as $=13$ and for humpback or triangular sections 15. (*Vide* fig. 50.)

For best cast iron $s = 5000$, which will mean a factor of safety of about 7.

For Admiralty bronze $s = 7000$.
  „ cast steel $s = 12,000$.
  „ best zinc bronze $s = 12,500$.

Or, assuming that a bar 1 inch square will carry in safety a B.M. of 1000 inch pounds, if made of good cast iron,

  1400 inch-pounds if made of Admiralty bronze,
  2400   „   „   „ cast steel,
  2500   „   „   „ good zinc bronzes.

Then

$$t = \sqrt{\frac{\text{B.M.}}{b \times y \times k}} \qquad . \qquad (2)$$

Here y is a factor of 0·45 for semi-elliptical sections and 0·40 for triangular.

The problem, however, may be approached in another and perhaps simpler way by taking the diameter of the screw shaft as the measure

DETAILS OF SCREW PROPELLERS.

of the work transmitted to the screw and deducing therefrom the forces set up by its blades. The capacity for resistance of the shaft to torque is the measure of the maximum-resisting couple of the propeller, and the turning moment about the axis is equal at any time to the bending moment then set up in the blades on their roots.

But it must not be overlooked that the screw shaft itself has to withstand the bending action due to the weight and inertia of the screw in addition to the torque of the engine, and as that means that the metal of the screw shaft is subject, among others, to alternating stresses rapidly applied, and in a sea-way these are very severe, its diameter is therefore considerably larger than it would be if subject to torque only.

But as the propeller itself has to bear shock and the stresses that arise from such external forces as come into play in a heavy sea, it will be as well, and certainly on the safe side, if the screw shaft itself be taken as the measure of the forces acting on the screw. So d will be taken as the diameter at the larger end of the taper end of the solid screw shaft, D the diameter of the screw, H the diameter of the boss, B the diameter of the flange of blades. Then—

When there are four loose blades . . $B = 0.66 \times H$
When there are three ,, ,, $B = 0.80 \times H$
Or, *vice-versâ*, when there are four loose blades $H = B \div 0.66$
,, ,, ,, three ,, ,, $H = B \div 0.80$

The thickness of blades at axis is the most important dimension of all, for, generally speaking, most of the others are mainly regulated by this one.

The torque necessary to damage a steel shaft whose diameter is d, and f the limit of elasticity, is

$$T = \frac{\pi d^3 f}{16} = \frac{d^3 f}{5\cdot 1}.$$

For mild steel subject to shear, the limit of elasticity may be taken at 27,000 lbs.; substituting this value for f

$$T = d^3 \times 5300.$$

Now this means that a torque T will, if constantly applied, quickly destroy the shaft of a diameter d inches.

If L be the maximum total load on the screw, having n blades, the load on each is $\frac{L}{n}$, and if the centre or resultant of the forces composing the load be applied at a distance Y from the axis, the resistance to turning is $L \times Y$, and at danger point $= d^3 \times 5300$. If it be supposed that the blade root is at the axis, the section will be in line with the axis; and suppose it to be in shape a segment of a circle.

If t be its middle thickness and b the breadth of blade, its power to resist bending will vary as b and as t^2, so that $L \times Y = nbt^2 \times F$.

But $L \times Y = d^3 \times 5300$; then $n \times b \times t^2 \times F = d^3 \times 5300$;

or
$$n \times b \times t^2 = d^3 \times \frac{5300}{F},$$

and
$$t = \sqrt{\frac{d^3}{n \times b} \times \frac{5300}{F}}.$$

Now, substituting E for $\frac{5300}{F}$.

thickness of blade at axis $= \sqrt{\frac{d^3}{n \times b} \times E}$.

The value of F has been calculated by taking the elastic limit for the zinc and strong bronzes and cast steel and 5 tons per square inch for the cast iron, and assuming the section to be a segment of a circle.

Value of F is 1400 for Admiralty bronze, and . E is 3·80
   ,,   ,, 2600 ,, manganese best and zinc . E is 2·04
   ,,   ,, 2500 ,, phosphor bronze   E is 2·12
   ,,   ,, 2400 ,, cast steel . . E is 2·21
   ,,   ,, 1000 ,, cast iron . E is 5·3

N.B.—For a smoothwater or summer service or for screws seldom worked at full power E may be reduced by 20 to 30 per cent.

Example 1.—What should be the axis thickness of the three blades of manganese bronze for a shaft 10 inches in diameter and a boss 25 inches long?

$$\text{Thickness} = \sqrt{\frac{1000}{3 \times 25} \times 2·04} = \sqrt{27·2} = 5·22 \text{ inches.}$$

Thickness of Blade at Root.—The thickness at a distance equal to $1\frac{1}{2} \times d$ from axis $= \frac{0·5 \times D - 1·5d}{0·5 \times D} \times t$;

and as approximately $d = D \div 12$,

thickness at root $= 0·75 \sqrt{\frac{d^3}{n \times b} \times E}$.

Example.—What should be the thickness at the root of the blades of a cast-iron screw with four blades and a boss 42 inches long? The shaft is 18 inches in diameter.

$$\text{Thickness} = 0\cdot 75 \sqrt{\frac{18^3}{4 \times 42} \times 5\cdot 3} = 0\cdot 75 \sqrt{184} = 10\cdot 17 \text{ inches.}$$

The thickness at tip should be one-fifth that at root; that is, thickness at tip $= 0\cdot 15 \times t$.

Thickness through the centre section radially should be calculated at one or two points as a check on those obtained in the usual way by drawing a line from the point E at a distance CE equal to t from the centre to the tip A, making thickness at $A = 0\cdot 15 \times t$, and lying a curve so as to make the line AE a tangent to it. In the case of blades broad at the middle portion, as in fig. 51, the curve may come within AE as shown.

Shape of transverse sections of a screw blade should be "ship-shape," and designed so as to pass through the water with least resistance (*vide* fig. 50). For this purpose the maximum thickness or, as it were, the greatest beam, should be nearer the leading edge than the following; that is, the "fore-body" is comparatively shorter than the "after-body," and should be in the proportion of 3 to 7, if possible; for, judging by experiments with torpedoes, a long after-body is more necessary for high speed under water than a fine entrance. It remains yet to be seen if better results would not be got from true ship sections than by the present half elliptical sections necessary to get the flat acting surface. Such sections certainly would be better near the root of the propeller, as shown in fig. 52, where the plain line section BHAE is as made in accordance with practice; the dotted line section BKAM is formed with the ordinary face line AEB as a centre and the thickness at corresponding spots the same. The part shown by the dash line HFE is as often done with large solid screws to do away with the shoulder which the back of the blade presents to the water on the passage of the blade through it.

All sections near the boss may with advantage be formed so that the leading part may present a wedge form in the direction of motion instead of the round shoulder at the back of the blade. (See fig. 50.)

The dimensions as above obtained may be relied on to give a propeller which shall be in accordance with good practice and have

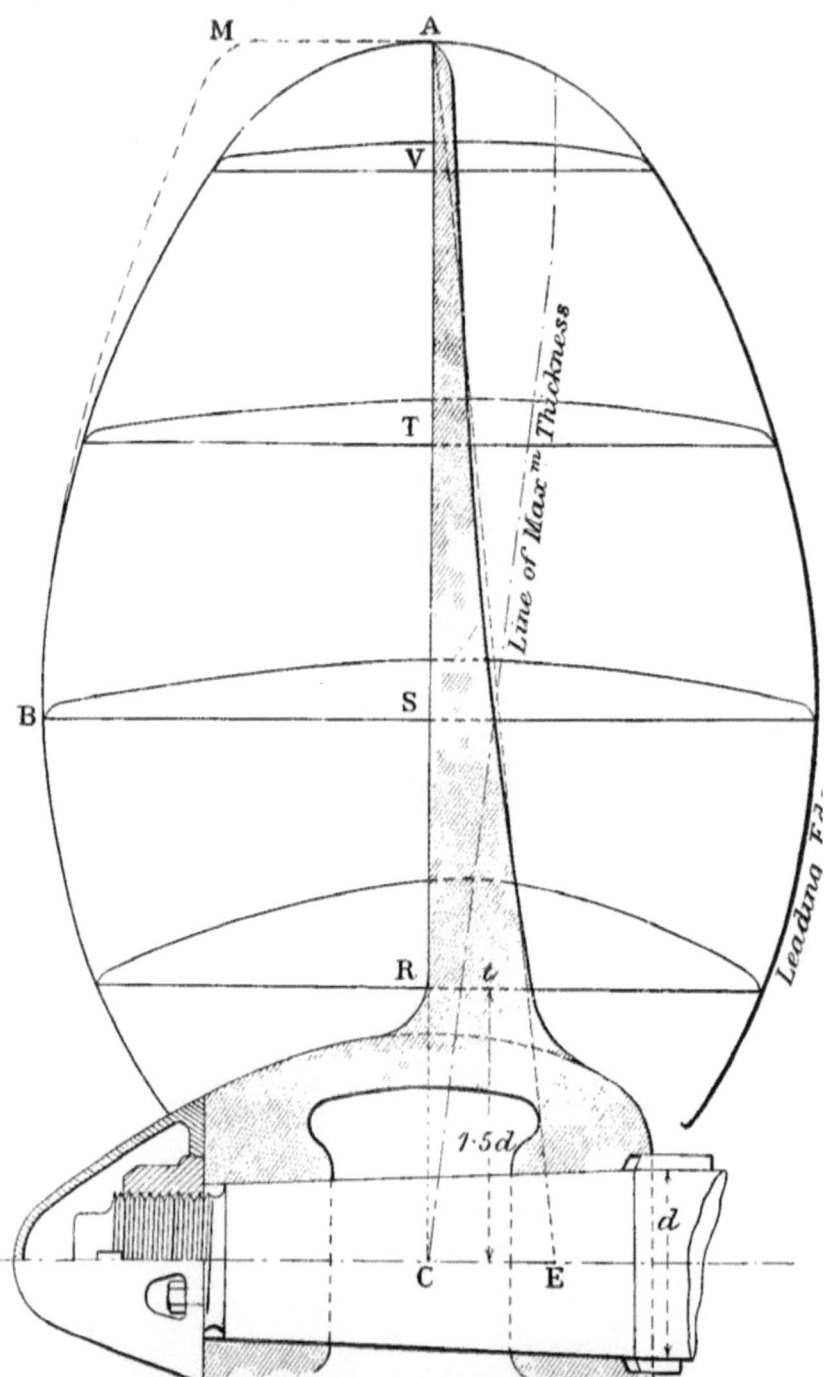

Fig. 51.—Solid Cast-Iron Propeller Blade.

DETAILS OF SCREW PROPELLERS. 179

practically the same factor of safety as the shaft. Full advantage, however, cannot be taken of the strength of the strong bronzes, inasmuch as the blades, while not breaking, would deflect so much as to detract from the efficiency of the screw. Hence for Atlantic work the thickness given by these rules should not be reduced; but, on the other hand, as already said, the value of E may be decreased by 20 to 30 per cent. for smooth-water ships or those using their full power but seldom.

It may be well also to remind designers that blades even of zinc bronze do sometimes disappear and the engines may, especially in twin- and triple-screw ships, continue to transmit full power to the remaining blades for some time before the casualty is appreciated.

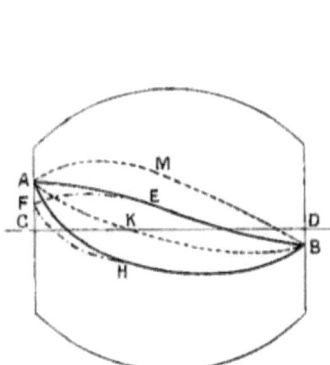

FIG. 52.—Various Root Sections of a Screw Blade.

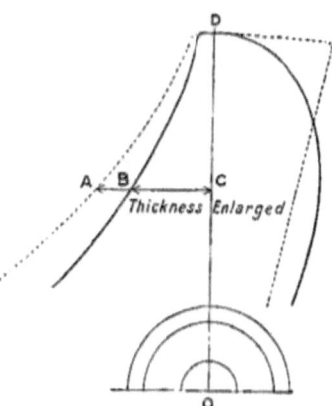

FIG. 53.—Longitudinal Sections of Typical Screws of Equal Area of Blade

A little allowance for this contingency, even if remote, may be made and looked on as so much insurance premium.

Fig. 53 shows the true shape of the longitudinal section of a screw blade if it is to have an uniform stress throughout the dotted line for a fantail and the plain line for an oval form of the same area. It will be found that in ordinary practice blades have the weakest part near the root, since they usually break there. It would therefore be better to take metal from the middle and add it to the root in a general way as shown in these sections AC, BC, which are drawn to a larger scale horizontally than the blades to make its shape and curvature more clear; it is common practice to draw them of the full thickness.

If the size of the shaft has been arbitrarily fixed on, as is the case in the Navy, when the shafts are hollow and enlarged to give stiffness to avoid vibration between the bearings, then d must be calculated by the following rules.

Diameter of Screw Shaft.—If this is not given, or is otherwise not available, and if p is the boiler pressure,

$$d = \sqrt[3]{\frac{\text{I.H.P.}}{\text{revolutions}}} \times R.$$

Two-cylinder two-crank compound engines, $K = 12 \sqrt{p}$.
Three-cylinder three-crank compound engines, $K = 9.8 \sqrt{p}$.
Three-cylinder three-crank triple-expansion engines, $K = 7.3 \sqrt{p}$.
Four-cylinder four-crank triple-expansion engines, $K = 8.0 \sqrt{p}$.
Four-cylinder four-crank quadruple expansion engines, $K = 8.0 \sqrt{p}$.

When the boss and blades are in one casting, as shown in fig. 51, the thickness of metal x around the borehole should be determined as follows:—

$$2d \times x \times f \times \left(d + \frac{d}{7}\right) = d^3 \times 5300;$$

$$x = d \times \frac{5300}{f} \times \frac{3}{7} = d \times \frac{2271}{f}.$$

For cast iron $f = 8000$ and $\frac{2271}{f} = 0.284$.

For cast steel $f = 11,500$ and $\frac{2271}{f} = 0.197$.

For Admiralty bronze $f = 10,000$ and $\frac{2271}{f} = 0.227$.

For manganese and zinc $f = 12,500$ and $\frac{2271}{f} = 0.181$.

For phosphor bronze $f = 13,000$ and $\frac{2271}{f} = 0.175$.

Thickness of boss, fore end $= 1.0 \times x$.
,, ,, after ,, $= 0.85 \times x$.
,, ,, outer shell $= 0.74 \times x$.

Solid screw *versus* **loose blades** is a subject that has been well thrashed out in the past. The efficiency of the solid screw in the mercantile marine was always, as seemed to be, higher than that of a screw with loose blades; and so it was in nine cases out of ten, for the blades of the latter were attached to the boss by means of a flange standing out from the boss and surmounted with the nuts of the studs and sometimes the stud ends. No attempt was made to shield them, or to mimimise the resistance. Then, too, the blades were often of steel and quite out of pitch, besides being rough on the surface, whereas the solid screw was made of cast iron smoothly cast, with the blades very fairly alike in shape and pitch; the boss an ellipsoid without obstructions, and the blade roots "faired" to the boss. Some of the early naval four-bladed screws were rather rough and ready about the boss; to-day they are as neat and well finished as the others.

The loose-bladed screw as fitted in H.M. ships, and also in high-class express steamers, are as efficient as any solid ones, and possess the advantages that the loss of a blade does not condemn the whole screw, as a spare blade can be fitted by a diver of experience, so that to refit it is not necessary always to go into dry dock. With the ordinary cargo steamer, whose "light" line is generally below the screw boss, there is no need even of a diver, for all the blades can be examined and, if need be, changed without taking the ship out of water. If, however, a diver cannot be obtained and there is no dry dock convenient, the ship sometimes can be "tipped" so as to raise the boss sufficiently near the surface to get a blade off. With loose blades the screw may have the boss of cast iron or cast steel, while the blades are of the very best metal and make; while the solid one must be all of one metal throughout, thereby adding to the cost if a costly metal is used. On the other hand, when a ship has been a long round voyage, she must go into dry dock or on a slipway to have her bottom cleaned and repainted. The opportunity then arises to examine the stern-shaft and its bearings; this necessitates the taking off of the propeller, which is also carefully examined, whether it be a solid or loose-bladed screw.

When a screw has loose blades they are often made of a different metal from the boss. Formerly in the Navy the boss was always made of Admiralty bronze, while the blades were sometimes of one of the other bronzes. At present the boss may be of any approved bronze. In the mercantile marine blades and boss were made of cast iron; with the advent of steel blades the cast-iron bosses still remained until some of the larger ones split from time to time; and as the steel makers could make steel bosses, the practice to have blades and boss of that metal became a rule. To keep down cost when the mercantile marine began to use bronze blades, the bosses continued to be of steel or cast iron, except in special cases when both were of bronze, the boss being of a cheaper variety of metal. In a general way there is no reason why the boss should not be of steel on a steel ship or even of cast iron on smaller ships, except that they corrode somewhat from the galvanic action unless care is taken to prevent it.

The earlier way of fitting blades to the boss was by forming a shank through which a cotter could be driven, as shown in fig. 38. So long as there were only two blades this method was possible and good for bronze screws. With cast-iron blades it was necessary to have a flange formed at the base of the blade through which screw bolts were fitted, securing it to the boss. With large screws the strain on the bolts was very heavy, and blades sometimes broke away from the boss. It was then found necessary to bed them into the boss either by a central spigot or by dropping the flange itself into a recess formed in the boss to fit it, as shown in fig. 43. All shear was then taken from the bolts, and their function was almost limited to holding the blade down to its seat against centrifugal force; the front ones, however, do hold the blade against the thrust and resistance to turning.

The Admiralty engineers have always insisted on having the bolt heads recessed, so that when a cover plate is put over them the sphere is complete and the boss has no obstructions. But not so with the mercantile engineer; for he has no cover plates, even when the boss is spherical and the blades fitted in recesses; he also often prefers steel studs having capped nuts of bronze fitted to them so that no water can get to them. Finally, when every nut has been "hardened up" and the steel set-screws tightened, a covering of Portland cement will be put on the lot and the semblance of a sphere attempted with this plastic material.

It need hardly be said that the screws of express steamships, whether cross-channel or cross-ocean, are treated in a manner more approaching that of the Navy, and the more so since high-speed reciprocating engines and turbines have come into use.

The number of bolts in a blade flange should, as a rule, be an odd one, so that there may be one more on the acting side than at the back, as they are always under stress when going ahead, while those at the back take little or none of the thrust load then, and only come on full load when the engine is going

DETAILS OF SCREW PROPELLERS. 183

astern; now, as the power developed in stern gear in most ships never exceeds half the full-speed power, there is so need for so many, even then.

Consequently small screws have three and two studs or bolts to each blade, while the very largest screws have nine—five in front and four in rear.

As a rough guide the following may be taken:—

$$\text{Number of bolts to each blade} = \frac{\text{diameter shaft in inches} + 6}{3}$$

By this rule, with shafts up to 9 inches diameter there would be five bolts, above that size and up to 15 inches there would be seven bolts, and from 15 to 21 inches nine bolts.

The diameter of bolts when made of steel, manganese bronze, zinc bronze, or naval bronze can be determined by the following rule:—

$$\text{Diameter of bolts or studs} = \frac{d \times Z}{n}.$$

Where d is the diameter of shaft as before, n is the number of bolts to each blade and Z is 1·6 for a three-bladed screw, 1·3 for a four-bladed one.

Centrifugal force produces in the screw blade at all times some stress, and at high revolutions the stress becomes serious, so much so, in fact, that destruction of blades is due sometimes to this source with screws driven by turbines.

Within moderate velocities the forces set up by inertia really tend to balance those by hydraulic pressure on the blade. That is to say, that whereas the hydraulic action tends to bend the blade in a direction opposite to that of revolution, the inertia of the blade tends to make it bend the other way as well as to "throw off."

The forces acting on a screw blade due to its velocity can be calculated from the usual formula where W is the weight of a blade in pounds, r is the distance of its centre of gravity from the axis of rotation, g is gravity, and taken at 32, v the velocity in feet per second:—

Then
$$C = \frac{W}{g} \times v^2,$$

and the tension on the bolts $= \dfrac{W}{g} \times \dfrac{v^2}{r},$

$\frac{v^2}{r}$ being of course the accelerating force, and called usually the *centrifugal force*.

When a propeller is in motion on normal conditions running at R revolutions per minute,

$$v = \sqrt{\text{pitch}^2 + (2\pi r)^2} \times R \div 60.$$

As an example take the case of a screw propeller 12 feet diameter, 15 feet pitch, 200 revolutions per minute; centre of gravity of blade is 3·2 feet from the centre; it weighs 1600 lbs. Determine the bending moment on the root distant 1·8 feet from the *c.g.* and the tension on the screw bolts screwing it to the boss.

$$v = \sqrt{15^2 \times (2\pi \times 3\cdot 2)^2} \times 200 \div 60 = 84 \text{ feet per second.}$$

$$C = \frac{1600}{32} \times 84^2 = 352,800 \text{ lbs.}$$

Tension on bolts = 352,800 ÷ 3·2 = 110,125 lbs.
If seven bolts, tension on each = 15,732 lbs.
Bending moment due to C = 352,800 × 1·8 = 635,040 foot lbs.

This is, however, in a plane through the face at the *c.g.*, and therefore is resisted by the section at the root longitudinally.

Taking circular motion and no advance of the screw,

$$v = 2\pi \times 3\cdot 2 \times 200 \div 60 = 67 \text{ feet.}$$

Then
$$C = \frac{1600}{32} \times 67^2 = 224,500 \text{ lbs.}$$

The bending moment on a plane at right angles to axis = 224,500 × 1·8 = 404,100 ft. lbs.

Taking an extreme case of an Atlantic steamer driven by turbines so that each screw receives 18,000 I.H.P. at 180 revolutions, the diameter being 16·6 inches, the pitch 18 feet, the weight of each blade 11,200 lbs., its *c.g.* being 4·5 feet from the axis and 2·0 feet from the root.

Here velocity = $\sqrt{18^2 + (\pi 9)^2} \times 180 \div 60 = 111$ feet per second.

$$C = \frac{11,200}{32} \times \frac{111^2}{2240} = 1925 \text{ tons.}$$

Taking circular velocity only,

$$C = \frac{11,200}{32} \times \frac{85^2}{2240} = 1129 \text{ tons.}$$

Tension on bolts = $\frac{1129}{4\cdot 5} = 251$ tons.

DETAILS OF SCREW PROPELLERS. 185

If thirteen bolts to each blade, the load on each = 19·3 tons in addition to that due to the pressure on the blade.

The weight of a screw propeller can, of course, be calculated; but to do it accurately is a long and tedious job, and scarcely repays the trouble taken. It can be estimated with a fair degree of accuracy by the following formula:—

Weight of a complete screw propeller in cwts. =

$$\frac{\text{surface} \times \text{thickness at root}}{K},$$

surface being taken in square feet and thickness in inches.

When made solid of cast iron $K = 4\cdot5$
    ,,   ,,   bronze $K = 3\cdot8$
When fitted with separate boss and loose blades of cast iron $K = 3\cdot0$
  ,,   ,,   ,,   ,,   ,, steel $K = 2\cdot8$
  ,,   ,,   ,,   ,,   ,, bronze $K = 2\cdot5$

CHAPTER XIII.

GEOMETRY OF THE SCREW.

As already stated, the acting face of the ordinary screw is part of a true helix, and the traces of it on concentric cylinders having their axes common to it enable it to be drawn in plan, and from this plan elevations can be projected on planes parallel to and at right angles to the axis. In fig. 54 AB is the diameter of the screw to be drawn at any convenient scale of which C is the centre. CQ, at right angles to it, is half the pitch.

ADB is the half circle of tip, and GN, HM, JP are the ends of the other concentric cylindrical surfaces.

This semicircle ADB is divided into equal sectors FAC, EFC, etc., and CQ, into the same number of equal parts AX, XY, etc.

Project the point F on the base line AB and produce the line so as to cut the horizontal line through X at K.

The point of their intersection K is on the trace of the helix on the outer cylinder ABA'. Point L and all the other points on this trace can be found in the same way, and the spiral line ALOA' can be drawn through them.

In the same way the trace GOG' on the cylinder whose radius is GC can be drawn; as also HOH', JOJ', etc.

The developed surface of the blade as decided on should then be drawn with C as the centre; nn' is the portion of the blade intercepted by the cylinder HM so that the chord nn' is the breadth of the blade seen on looking down on that cylinder.

From the corresponding trace HOH cut off a portion hh' equal to nn'; and from the other traces cut off portions corresponding to those intercepted by the other circles GN, AD, etc.

Now draw a line on each side through the terminals of these cutoff portions and the outline of the blade in plan is made at hOh'.

Now, with Q as centre, draw the circles representing, as it were, the other ends of the cylinders; through h and h' draw lines parallel to CQ, then produce them at right angles to QA', cutting cylinder HM end at k and k'. Then k and k' are points on the end elevation

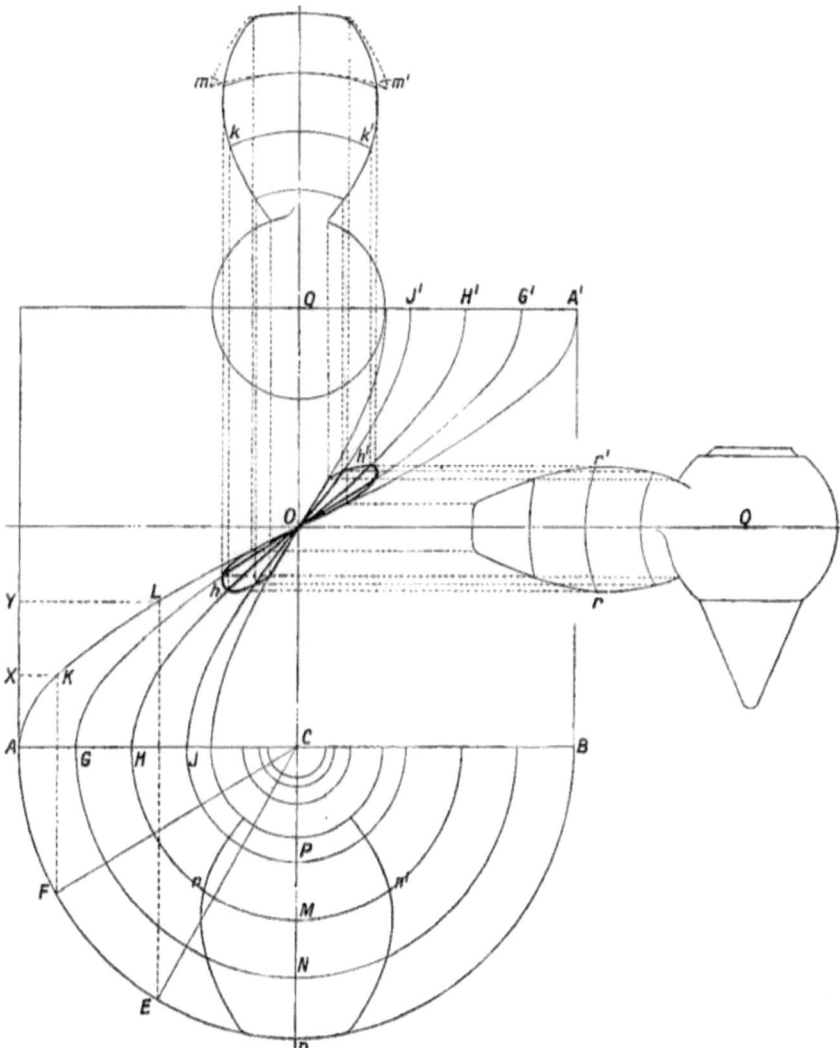

Fig. 54.—Method of Delineating a True Screw Accurately.

or projection of the screw. m and m' are found in the same way, as also any other points. Lines drawn through these points mk, etc., $m'k'$, etc., will be the outline of the screw as seen looking in line with its axis.

Now rr' is the longitudinal view of the trace on the cylinder HM, so that by drawing lines hr and $h'r'$ at right angles to the axis CQ, rr' will represent in side elevation the line hh'.

In a similar way all other points can be found; and lines drawn through them will give the outline in side view on the projection on a plane parallel to the axis of a blade whose developed form is $n\mathrm{D}n'$.

To show the projected width of the blade at mm' it is necessary to suppose the blade to be twisted round till that portion of the blade is parallel to the plane.

The points mm' will then remain on a line drawn through them, and qq' will be the terminals and equal to $n\mathrm{M}n'$ in length, but the curvature of qq' will not have its centre at Q as being part of an ellipse.

In everyday practice, and more especially when dealing with propellers whose blades are comparatively narrow, so that no violence is done to the design or calculation, the portion hh' is assumed to be a straight line. If the screws have very broad blades and accuracy is demanded, then the above methods must be observed.

The rough and ready way usually observed in many drawing offices is illustrated by fig. 55.

Hence AB is π × diameter at tip, and AD is the pitch, both drawn to quite a small scale, say $\frac{1}{4}$ inch to the foot for moderate size screws, and $\frac{1}{8}$ inch to the foot for very large ones.

Divide AB into eight equal parts at D, E, F, etc.

Join D, E, F, etc., to O and produce them beyond; then AOD is the angle of the blade at $\frac{1}{8}$ the diameter, AOG that at $\frac{1}{2}$ the diameter, etc., AOH that at $\frac{5}{8}$. Draw D_1MLM as the developed acting surface of the proposed screw.

The breadth on circle H_1 is MM_1. Cut off a portion mm of HOM equal to MM_1. Project this back on circle H_1, so as to intercept NN_1. Carry out this operation at each of the other circles LKJ, etc. The result will be that on drawing lines through the terminals, the projected surface D_1NLN_1 will be formed, which is that of the screw blade on a plane at right angles to the axis.

By projecting from the plan horizontally to a series of circles $D_{11}E_{11}F_{11}$, etc., in the same manner the surface is projected to a plane parallel to the axis. It is, of course, understood that to be correct MNN_1M_1 should be all in one horizontal line, because point N is supposed to be the position of M when the blade is twisted to pitch from a transverse position.

Varying pitch may mean that in a complete revolution the pitch has changed from x to y; so that, at one edge, the entering angle is that due to a pitch x; while, at the other or discharging

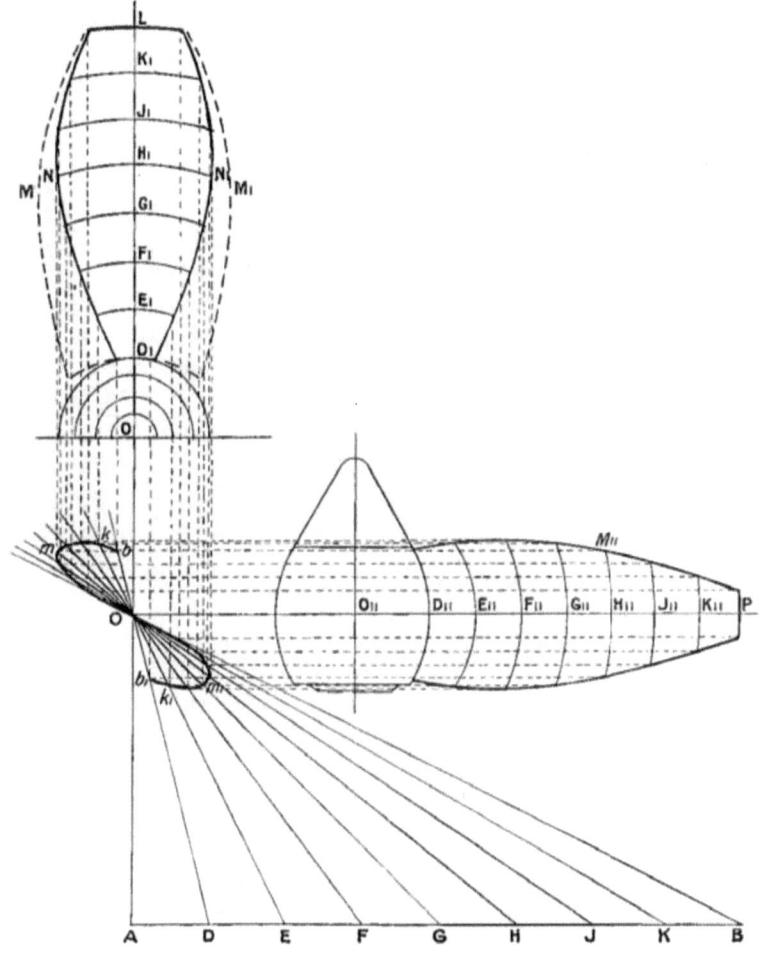

Fig. 55.—Simple Method of Delineating a Screw.

angle, it is that due to a pitch y. Woodcroft's screws, when of one convolution, were formed in this way. The method by which to draw the trace of that screw is shown in fig. 29, and described on page 106.

Through the points APS, YXW, the curve drawn is the trace of the blade tip of a screw of one convolution of a screw whose pitch varies from BL to BC as developed on the enwrapping plane.

In practice, with screws having a length of from ⅛ to ⅓ the pitch, it was, and is still, usual to form a leading portion of the blade, say from one quarter to one half of the breadth, with a pitch corresponding to a speed slightly greater than that expected of the ship, and the remainder greater and sufficient to give the necessary acceleration.

The blade face can then be drawn as shown in fig. 56, where AB represents π × diameter and BD the pitch at entry, and BC the pitch at delivery.

Join AC and AD, when the angle of the blade at entry is BAD and at delivery BAC.

Produce AC and cut off AF and AE in the proportion decided on. Then apply a curve so that AE and AF are tangents to it and the line EAF is typical of the blade surface.

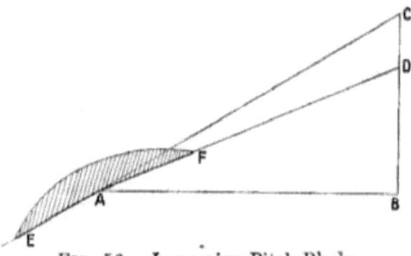

FIG. 56.—Increasing Pitch Blade.

Screws are generally made with a rounded back to the blade section, so that it is formed from a part of an ellipse, or an ellipse-like curve as shown in fig. 50. But sometimes the back is practically a pair of planes inclined to one another, also as shown in fig. 50.

It is claimed for the latter that a stiffer blade is obtained with the same weight of metal as in the round backed; and, further, that it offers less resistance in its passage through the water, especially at high revolutions. When formed in this latter way it is usual to make the maximum thickness slightly greater. So that in formula (1) (see p. 174) the constant 13 becomes 15; and 0·45 becomes 0·40 in formula (2).

Hollow Castings.—Messrs J. Stone & Co. have for some time been in the habit of making the root end of large propeller blades hollow. A careful moulder can do this with little or no risk of a waster, or even of a faulty casting; and as the metal by slow cooling at the very thick part is always highly granular and often spongy, the withdrawal of it means really no loss of strength, while it does mean less weight and cost.

Root Sections.—It is not an uncommon thing for the leading edge near the root to be shifted from A to F so that the section there does not present a broad round shoulder in the direction of motion (E to A) as shown by AH, fig. 52, p. 179.

The suggestion that propellers would be better made so that what is now their acting surface, as at AEB, should become the middle of the blade section, is well worth a trial on a ship of fairly good size.

In fig. 52 BKAM would be such a section at the root of the screw instead of BHAE as at present, or of BHFE, as has sometimes been adopted.

To found a propeller the helical surface is formed on a loam facing to a rough brick foundation, or it may be on common green sand, by a strickle board having an edge of metal formed to the shape of the acting face of a longitudinal section of the proposed screw and made to travel on spiral guides, one just beyond the tip and the other near the boss, while it revolves around a fixed bar standing at right angles to the plane or the flat cast-iron bed on which the mould is built. The spiral guides are made of triangular pieces of plate, whose bases to height are in the same proportion to the circumference of the circle on which they stand to the pitch of the screw; they are bent to a portion of that circle. When the bed is moderately dry, a centre line is inscribed on it corresponding to the centre of the blade. The blade breadths are then set out on inscribed circular lines, and to their extremes is applied a piece of square bar iron bent to shape so as to lie evenly on the mould. Lines are then inscribed from it and the blade surface form is completed. Parting sand, quite dry, is then sprinkled on the mould face, and pieces of thin wood cut to the shape of the sections at various points are laid on the face and the spaces filled in with "black" sand; while over the whole a coating of loam or green sand is laid and the mould completed in the usual way. When dried or baked, as the case may be, the moulds are opened and the black sand and sections removed, and after cleaning up, fairing, and rubbing smooth, the surfaces to be exposed to the molten metal are coated with "blacking" or plumbago mixture; they are again put together, carefully set, and otherwise dealt with as other moulds are.

If a blade only is required, as in the case of a loose-bladed screw, a flange is fixed at the end of a blade mould of the form, etc., suitable to it constructed as above. If it is a solid screw there are wing boxes, one for each blade, and they are set at the proper angles with a boss mould in the middle of them.

Pattern Blades.—In old days it was a common and costly practice to make a pattern blade in mahogany, and as samples of pattern-making they were very beautiful; but by the method above described almost as good a pattern can be obtained in cast iron, and certainly a cheaper and more durable one.

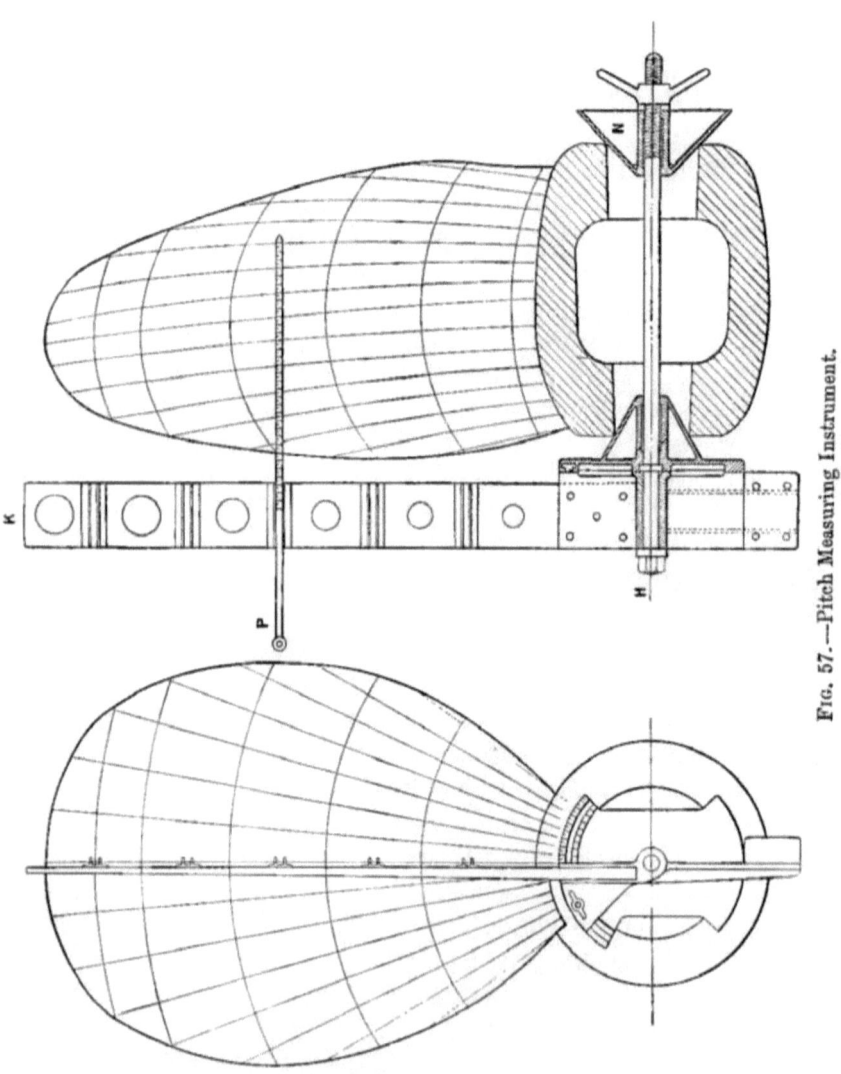

Fig. 57.—Pitch Measuring Instrument.

Having a pattern, any number of blades may be made from it in dry sand. Great care, however, is necessary in the pattern-making as well as in moulding to ensure getting quite true and trustworthy blades and thus to avoid having to chip, file, and grind the acting

GEOMETRY OF THE SCREW.

surfaces to get the best performance on high revolution trials, all costly and slow and unsatisfactory processes, and often ending in having to accept a pitch different from the exact one designed rather than condemn an otherwise good screw to the scrap heap.

Methods of ascertaining the pitch of a propeller have been from time to time devised, and special instruments made whereby the approximate pitch is seen at a glance; but none can exceed in simplicity, accuracy and usefulness the apparatus shown in fig. 57. The radial batten is of dry, well-seasoned wood which is practically unaffected by atmospheric conditions, such, for example, as mahogany or cedar. The metallic fittings may be made, with advantage, of aluminium, now that metal is so cheap (£70 per ton). The dial plate is marked off in degrees, and each angular movement usually is 5 degrees. The horizontal staff or measuring rod is of mahogany, fitted with a metallic tip so that it touches the blade face in line with the edge next the batten and capable of sliding in the guides on the batten. It can be with advantage marked in inches and fractions of an inch, but if the angular movement is always 5 degrees the staff may be marked with one-sixth of an inch as the unit, as then each unit for that angular movement means a foot of pitch. A useful addition to the installation is a metallic scale marked on the four edges with units to suit four different angular movements and arranged to be secured to the measuring rod at the part where it slides in the guides attached to the batten. A reference mark or score is on each guide for this purpose.

CHAPTER XIV.

MATERIALS USED IN THE CONSTRUCTION OF THE SCREW PROPELLER.

THE very earliest of the screw propellers were often made of oak staves cut to shape and through-bolted to one another so as to have the appearance of a double spiral staircase. The "steps" at front and back were, however, "dubbed" away down to the angles so as to form a true helical surface. The patterns for the early screws were also made in this way, for there was then formed a two-bladed screw like those used by Smith and Woodcroft. To employ oak or elm timber was of course quite a natural thing to do, seeing that the floats of the paddle wheel, as also the big sweeps or oars, even then often used in sailing vessels as propellers, were of such wood.

The frictional resistance of the surface of these hard woods would be enough to condemn their use, even had not the inherent weakness of their structure already done so, as soon as the diameter of screws was increased beyond the model limits. One good service was rendered by the wooden screw, however, before it was put on one side, when a portion of Smith's early ones broke off, the surface was so reduced as to permit of greater revolution and more speed, thereby teaching engineers that a screw could have too much area of blade, as well as to demonstrate the magnitude of the evil.

Naval screws of bronze was the rule from the outset, and this was no doubt due in no small measure to the fact that naval ships were in the early days of the screw built of wood, copper fastened, and always sheathed with copper. Even to-day ships intended for the parts of the world which have inadequate dry dock accommodation are sheathed with teak planking, on to which the copper sheets are nailed, as they were to the old wooden ships. In the 'sixties of last century, when iron ships were taking the place of wooden ones for naval purposes, the practice of fitting bronze screws still obtained; and although fears of corrosion of the ships' plating were often expressed, they were never realised sufficiently to condemn the practice. As a matter of fact, there was sometimes corrosion in the immediate neighbourhood of the screw race, which, in some few cases, took the form of pitting and developing pustules under the red lead paint.

The fitting of a few slabs of zinc in metallic contact with the hull is, however, a sure preventative of corrosion.

The zinc bronzes and manganese bronze do not affect iron in the same degree, so that when the boss as well as the screw blades are of that metal there is no need for the protecting zinc slabs.

The Admiralty have continued to use "bronze" with very few exceptions; but there is not now the same limitation in proportions or ingredients that formerly obtained; for whereas then the "bronze" or "gun-metal" meant a composition of best selected copper with about 10 per cent. of tin, now it may have in addition to tin, zinc and some other metals, and in some cases be even without tin.

The following are the metals, or rather alloys, called "bronze," used for propeller blades, etc., in H.M. Navy, and also largely so in the mercantile marine for express steamships where safety and efficiency are carefully studied.

Admiralty bronze as used for propeller blades, bosses, etc., in the Navy is now a mixture of—Copper, 87 per cent.; tin, 8; zinc, 5.

When carefully alloyed and cast the ultimate tensile strength should be 15 tons per square inch with test bars. The average strength of pieces cut from castings should not be less than $13\frac{1}{2}$ tons with a stretch of $7\frac{1}{2}$ per cent. in 2 inches. Its specific gravity is 8·66.

Phosphor bronze, a metal sometimes used for the blades of ships, is an alloy of copper, 82·2 per cent.; tin, 12·95; lead, 4·28; phosphorus, 0·52.

It is harder, closer grained, and stronger than Admiralty bronze; its ultimate strength is 17 tons and sometimes higher per square inch; but the most important characteristic of this metal is that its elastic limit is very high; in fact, higher than the ultimate strength of Admiralty bronze. It has also a high resistance to shear. It is therefore very suitable for blades, and especially for the bolts used to secure the loose blades.

Aluminium bronze, containing copper, 95 per cent.; aluminium, 5·0, possesses great tensile strength even in the cast state, amounting to as much as 25 tons per square inch with an elongation of 42 per cent., while that with 7·5 per cent. aluminium and 2·0 per cent. of silicon has a still higher tensile strength and an elongation of 25·6 per cent. It has been used for propellers, but only to a small extent, as in cost and physical qualities it cannot compete with what are called zinc bronzes or with manganese bronze.

Manganese bronze, one of the zinc bronzes, was very generally used for propeller blades; it was discovered and perfected by the late Mr Percival M. Parsons, and castings of all sizes were made by

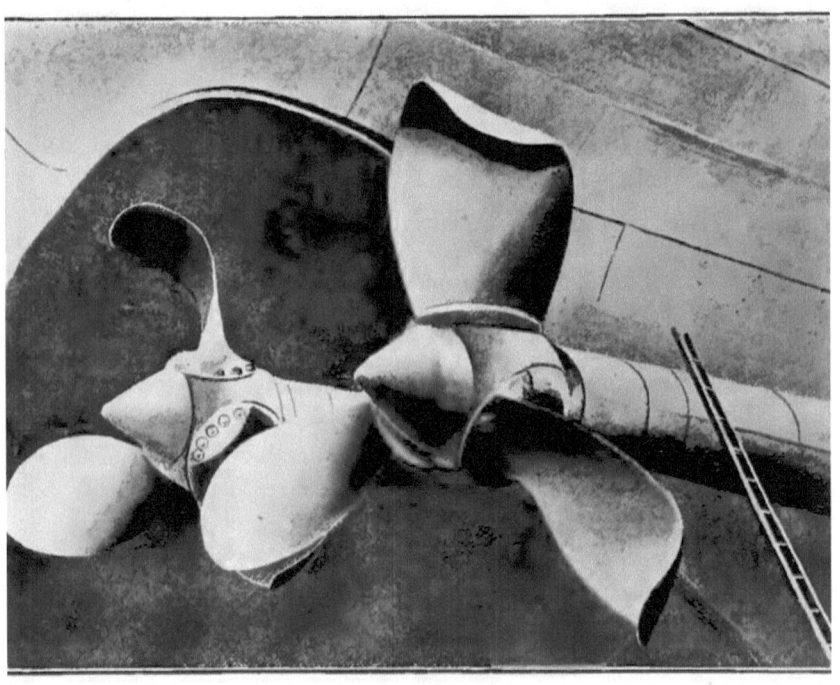

S.S. "Norman."

Fig. 58.—In Dry Dock, after stranding on Coast of Africa, showing bent Propeller Blades of Parson's Manganese Bronze.

him, and are now by the company he founded. It has an ultimate strength of 30 tons per square inch with an elongation of 21·5 per cent., and is tough. It makes very good bolts, and withstands the action of sea-water most successfully. The first blades were made in 1879, and the screw of H.M.S "Colossus," after twenty-four years' use, showed no signs of corrosion.

The Parsons Manganese Bronze Company now make a special bronze for screw propellers which possesses in a high degree the qualities necessary for a successful screw blade, and the extent of

MATERIALS USED IN CONSTRUCTION OF SCREW PROPELLER. 197

punishment such screws can successfully withstand may be seen by examining figs. 58, 60.

Stone's bronze is now one of the best known of the "zinc bronzes," which are so called because of the large proportion of zinc in their composition. This alloy has in ingots and castings of any size, as supplied by J. Stone & Co., copper, 56·1 per cent.; zinc, 40·6; tin, 1·05; iron, 1·67; lead, 0·47.

Messrs Stone's No. 3 bronze is a very remarkable alloy, having physical properties which make it very suitable for such things as propeller blades.

It has the very high ultimate strength of 35 tons per square inch with an elongation of over 30 per cent. This metal, therefore, is exceedingly tough, and when a blade receives a heavy blow, it generally bends without breaking, and often can be re-set on being heated and carefully handled (fig. 59).

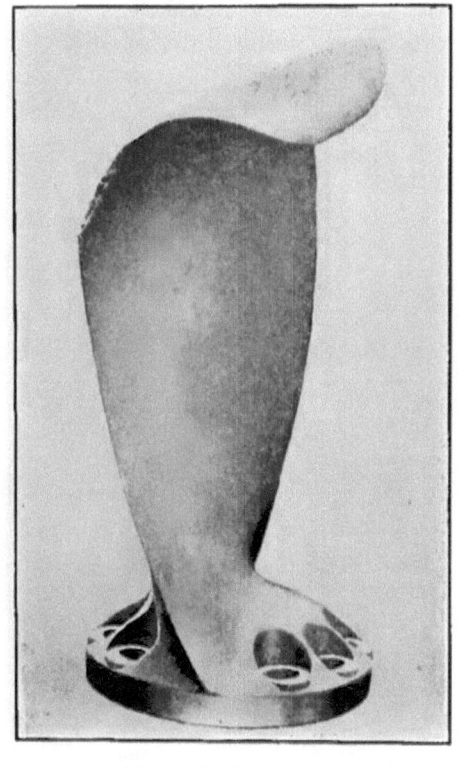

FIG. 59.—Blade made of Stone's Bronze, and bent by accident.

A bar of this metal an inch square placed on supports 12 inches apart will sustain a load at its centre of 5750 lbs. without fracture, with a deflection of 1·01 inches; 6820 lbs. were required to break it down to 2·32 inches.

The following figures show the progress with similar bars cast in sand from crucibles :—

TABLE XVII.

Load, lbs.	3000	3500	4000	4500	5000	5500	6000	6350
Deflection, inches . .	0·18	0·26	0·39	0·57	0·82	1·18	1·65	Rupt.

Bars of the same size and length—

If of cast iron, give way at about 2000 lbs.
  „ gun-metal,   „   „   „   2600 to 3000 lbs.
  „ cast steel,   „   „   „   6100 lbs.

The elastic limit of No. 3 is 17·05 tons, or 47·3 per cent. of the ultimate strength, viz. 35 tons.

Tests made by Professor A. K. Huntingdon on the resistance to torsion of this metal are interesting. A bar $\frac{7}{8}$ths inch diameter and $5\frac{1}{2}$ inches long was subjected to twisting, with the following results:—

TABLE XVIII.

Torque, Inch lbs.	1195	2170	3250	4340	5425	6510	6980
Angular movement, degrees	3·60	6·84	23·76	64·05	133·9	222·1	Rupt.

A bar of this metal $\frac{1}{2}$ inch in diameter and 1 inch long will withstand compression up to 20,000 lbs. with a reduction in length of 0·174 inches. With a load of 5000 lbs. it is only 0·004 inch. It took 25,000 lbs. to break it, or nearly 56 tons per square inch. It may, therefore, be stressed in compression up to 11 tons per square inch under test with safety.

Naval brass is an alloy of copper, 62·0 per cent.; zinc, 37·0; tin, 1·0, and is a strong, tough metal often used for studs and bolts. It can be rolled and forged, and has an ultimate tensile strength of 24 tons per square inch with a stretch of 10 per cent. in two inches. Its elastic limit is about 42,000 lbs.; its resistance to shear, however, is only 55 per cent. of the tensile strength.

Bull's metal is very like Stone's in composition, but has no iron; its character is good.

Delta metal is another of the zinc bronzes; it has, however, 2·0 per cent. of manganese, but no tin. It also has a high tensile strength, and is tough.

Sterro metal, the earliest of these bronzes, consists of iron, 1·5 per cent.; zinc, 38·1; copper, 60·0; and has similar qualities to the others.

But all these zinc bronzes, while possessing very high ultimate strengths, have an elastic limit of only about 14 to 15 tons per square inch, and their resistance to shear is also comparatively low. They are, however, comparatively cheap and easy to machine. They can be smithed when red hot and be re-melted, but great care must be taken in smithing, as is also the case in melting both new and old metals, in order to get the best results. There is no secret in their composition, as it is known and has often been published; their good physical qualities depend rather on the mixing, melting, and general heat treatment than the exact quantities of the ingredients.

Messrs J. Stone & Co., the Parsons Manganese Bronze Co., and Billington & Newton, have made a special study of these bronzes; this, together with the great experience they have had in the moulding and casting of blades, ensured their turning out and supplying a thoroughly reliable propeller which has done much to popularise the adoption and general use of bronze blades, in spite of their high cost compared with that of cast iron.

Cast iron was the material of most of the screws of the mercantile marine from the earliest days, and has continued to be largely used even to-day. It is, of course, the cheapest of the metals and the easiest of manipulation, and can be got of fairly good quality almost anywhere. It is claimed for it as a virtue that when it receives a serious or damaging blow it breaks clean off and produces no obstruction; for a propeller blade this is perhaps an advantage, especially in the case of the single-screw ships, for when struck a heavy blow from wreckage, it should part off and sink rather than bend out of place and perhaps prevent the propeller from turning round in the frame, thereby disabling the ship.

Admiralty bronze blades have been in the past quite accommodating in this way, and have performed the feat more than once of breaking off and disappearing. The manganese and zinc bronzes, on the other hand, do bend, and that very considerably, under a severe blow; and many a blade of these metals has suffered terrible distortion without breaking (see fig. 60); but it must be a very rare event for a bend to be so bad as to foul the stern or rudder post of a single screw, and impossible to foul the brackets of a twin-screw ship.

Good modern cast iron propellers can be obtained whose ultimate strength is 10 to 12 tons per square inch, and as its yield point is nearly the same, its strength will compare favourably with that of Admiralty bronze, whose yield point is only seven tons; but unfortunately cast iron does not stand sudden heavy or continuous shock so well, and does not bend at all.

But cast iron has two other characteristics which go far to condemn it as a metal suitable for propellers, especially for those moving at high velocity. It rusts or oxidises quickly when exposed to sea water, especially to the aerated sea water abounding at the backs of the blades; and when oxidised in this way its surface becomes very rough, and hence causes great frictional resistance after a few weeks' service. It is true the blades may be cast quite smooth and coated with enamel paint, and that when so done the skin resistance is no greater than that of a bronze screw; it is generally true, however, that in practice it is difficult, if not impossible, to keep the paint on the outer part near the tips, just where resistance is most serious, and still more so on the backs. Further, it is quite impossible to maintain the leading edges of the blades so thin and sharp as is the case with bronze ones, even when the founder manages to cast them so. And, moreover, from the iron blades being so much thicker than bronze, their resistance to turning is much greater. Some four-bladed cast-iron screws were tried by the Admiralty on H.M.S. "Victor Emmanuel" in 1853, but they were seldom used after.

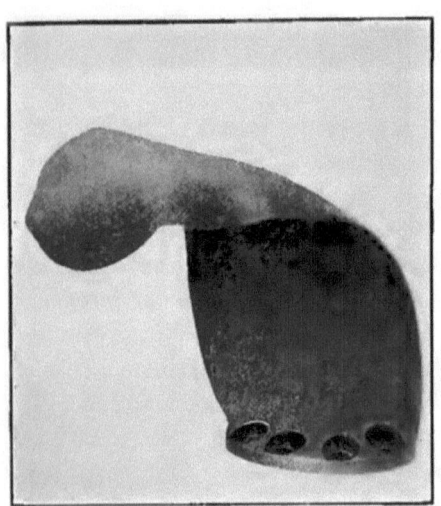

FIG. 60.—Manganese Bronze Co.'s Propeller Blade, bent, from s.s. "City of Paris."

Cast steel was introduced to the shipping world as a suitable material for large propeller blades by Messrs Vickers & Sons in 1870, and for many years that eminent firm supplied most of the large Atlantic liners with their products. The Admiralty gave this material a trial on a few ships whose bottoms were sheathed with zinc instead of copper. As, however, the zinc was not found to last long, its use was discontinued; and as steel did not agree with copper sheathing and bronze castings, blades of it were soon discarded from the Navy.

When first placed on the market, steel blades were very costly, almost as much so as bronze ones, especially when their extra weight was taken into consideration, together with the extra cost of machining such hard steel as was then used.

The cost of steel blades, however, came steadily down after about 1885, and eventually could be bought as castings at a price not very much higher than twice that of iron castings, consequent upon the use of the Siemens furnace for melting as against the crucibles of Vickers.

In spite of every precaution the steel blades corroded even worse than the cast iron ones had done; they were seldom cast of the form or pitch designed, and were nearly always of coarser pitch at the boss than the tip; in fact, no two blades were sufficiently alike for good working. Even those from the best makers were considerably inferior in the matter of correctness of pitch and being helical to the bronze or even to cast-iron ones. The machining and finishing of steel blades was, of course, always much more costly than bronze and cast iron, so that altogether, in the end, they were not really so cheap; and seeing that, as scrap, they were practically valueless, whereas the bronze blades were worth quite half their original cost, the latter gradually drove the steel out of the market. There are, of course, a considerable number still remaining in use, and steel founders work now more closely to the design than was formerly the case; they cannot, however, give the fine sharp edges of the bronze blades, or even equal the cast iron ones in this respect or in smoothness of surface, etc.

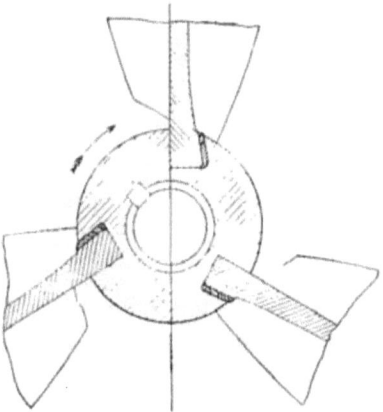

FIG. 61.—Blades Dovetailed into Forged Boss.

Forged steel blades have been made for the very high speed of revolution on torpedo boats and "catchers" of years gone by, when it was necessary to have blades as thin as possible consistent with strength and stiffness. It need hardly be said that such propellers were very costly, and as liable to corrosion as cast steel and cast iron. When so made they were often fitted to a wrought steel boss by means of grooves planed in the latter and held by wedges, as shown in fig. 61. Bronze blades were also fitted in this way to steel bosses, as such blades could then be cast very accurately, and any want of accuracy was removed by grinding. Moreover, with such a boss quite small propellers may have detachable blades.

CHAPTER XV.

TRIALS OF S.S. "ARCHIMEDES" AND H.M.S. "RATTLER."

The screw steamer "Archimedes," built for the Ship Propeller Co., a syndicate of F. P. Smith and his friends, was a sea-going ship designed by Pascoe and of the following dimensions: 106·7 feet long between perpendiculars, 21·8 feet extreme beam, and 13 feet depth of hold. Her tonnage was 237, draught 10 feet aft and 9 feet forward. She was propelled by two engines of 90 H.P., having two cylinders 37 inches diameter and a stroke of 3 feet. She cost £10,500, and in May 1839 steamed from Gravesend to Portsmouth in twenty hours. The following tables give particulars of her competitors and their performances on the Channel station between Dover and Calais in 1840.

Quickest passage from Dover to Calais was made by the "Archimedes," and took 1 hour 53½ minutes to perform.

Table XIX.—"Archimedes" and her Competitors in the Channel Trials.

Name of Ship.	Propeller.	Tonnage B.M.	Dimensions.		Area of Mid. Sec.	Engines N.H.P.	N.H.P. / Mid. Sec.	Speed on Trial.	Cylinders.
			Ft.	Ft.	Sq. Ft.				
"Ariel"	Paddle	152	108 long :	17·3 B.M.	95·0	60	0·632	10·4	..
"Beaver"	,,	128	102·2 ,,	16·0 ,,	84·0	62	0·738	11·2	..
"Swallow"	,,	133	107·6 ,,	14·8 ,,	84·0	70	0·833	10·4	..
"Widgeon"	,,	162	108·0 ,,	17·1 ,,	95·0	90	0·947	10·3	Two 39-inch diameter × 37 inches.
"Archimedes"	Screw	237	106·0 ,,	21·8 ,,	143·0	80	0·559	9·6	Two 37-inch diameter × 36 inches.

TABLE XX.—RESULTS OF TRIALS OF "ARCHIMEDES" AT DOVER, 1840

Name of Competitor.	Course.	Weather.	Conditions.	Speed of "Archimedes."	Winner by Time.
"Ariel"	Dover to Calais	Fair	Both using sails	Knots 9·75	"Archimedes" by 6 1
,,	Calais to Dover	,,	,,	9·75	,, ,, 5
"Beaver"	Dover to Ostend	,,	,,	9·50	,, ,, 4
,,	Ostend to Dover	,,	...	9·25	"Beaver" ,, 9
"Swallow"	To Calais and back	,,	Steam only	...	"Archimedes" ,, 1
"Widgeon"	19-mile course	Fine; fair wind	,,	8·50	"Widgeon" ,, 6
,,	,,	Head wind	,,	8·00	,, ,, 10
,,	,,	No wind	Quite smooth	...	,, ,, 3¼
,,	,,	Fresh breeze	Both using sails	...	"Archimedes" ,, 9

Admiralty Trials with H.M.S. "Rattler."—Experiments with the s.s. "Archimedes" having sufficiently satisfied the Admiralty that screws had such advantages for naval purposes as to warrant further trials, it was decided to build two sister ships and fit them with similar engines made by the same engineers, the one set geared 4 to 1 to the screw shaft of one ship, and the other coupled direct to a pair of paddle wheels in the other.

The "Rattler," a screw steamer, was 176·5 feet long between perpendiculars and 32·7 feet extreme beam and 15·5 mean draught of water. Her displacement was 1140 tons and the immersed midship section 348 square feet.

The "Alecto," the paddle steamer, was built to the same mould but differed somewhat at the stern from the "Rattler," as the latter ship had to be designed to suit the screw and its appurtenances (see fig. 7). Each ship had a set of Maudslay's twin cylinder engines having four cylinders 40 inches diameter and 48 inches stroke; the N.H.P. being 200 and the I.H.P. on trial 437 in the case of the "Rattler," when tried in January 1846 and running at 9·639 knots. The propeller was a Smith's common screw 10 feet in diameter and 11 feet pitch, with an acting surface of 22·8 square feet, running when at full speed 107·9 revolutions per minute.

In the competitive trials with the "Alecto" the ships were ballasted to a mean draught of 12·3 feet; the first run was between the Nore and Yarmouth Roads, under steam only, with the sea smooth and the weather calm. The time taken by the "Rattler" was 8 hours 34 minutes, and as the distance was 78¼ nautical miles, the mean speed was 9·2 knots. The "Alecto" took 8 hours 54 minutes, so that her mean speed was only 8·8 knots, but she attained that with only 281·2 I.H.P. as against the 334·6 of the "Rattler."

TABLE XXI.—ANALYSIS OF THE TRIALS OF VARIOUS SCREWS ON H.M.S. "RATTLER," 1843 TO 1845.

Designation of Screw.		1 Smith's Common.	2 Smith's Common.	3 Smith's Common.	4 Smith's Common.	5 Smith's Common.	6 Smith's Common.	7 Smith's Common.	8 Smith's Common.	9 Smith's Common.	10 Smith's Common.	11 Woodcroft Incg. Pitch.	12 Steinman's Screw.
Diameter of screw	ft.	9·0	9·0	9·0	9·0	9·0	9·0	10·0	10·0	10·0	10·0	9·0	10·0
Pitch		11·0	11·0	11·0	11·0	11·0	11·0	11·0	11·0	11·0	11·0	11·0	11·5
Length		4·25	3·0	2·25	1·71	1·00	1·17	3·0	2·0	1·5	1·25	11·55 / 1·7	
Number of blades		2	2	3	3	3	3	2	2	2	2	2	2
Acting surface	sq. ft.	60·6	42·8	52·3	38·4	36·0	27·0	51·28	31·16	25·6	22·8	28·0	
Pitch ratio		1·22	1·22	1·22	1·22	1·22	1·22	1·10	1·10	1·10	1·10	1·28	1·15
Surface ratio		·946	·670	·819	·634	·570	·422	·450	·398	·324	·298	0·439	
Revolutions per min.		94·0	106	94·3	92·0	98·3	103·4	98·0	100·6	102·1	103·6	107·5	100·9
Slip per cent. app.		22·4	24·4	32·5	18·8	21·8	16·0	12·9	13·4	12·2	10·4	27·7	15·73
Speed	knots	8·3	9·24	8·237	8·996	8·561	9·880	9·231	9·448	9·721	10·07	8·63	9·639
Indicated horse-power		350	395	351	341	364	401	360	372	378	383	398	373
Speed³ ÷ I.H.P.		1·643	1·990	1·563	1·559	1·723	2·404	2·182	2·266	2·428	2·666	1·613	2·398
Thrust, indicated	lbs	11,223	11,150	11,180	11,337	11,100	11,108	11,020	11,090	11,177	11,100	10,650	10,610
„ calculated		9,830	10,560	9,360	7,660	8,424	8,563	10,982	9,724	9,100	8,988	9,170	
Tr. resistance of ship		5,720	7,220	6,645	5,470	6,130	8,380	7,200	7,540	8,050	8,060	6,236	7,880
P. H.P.		251	299	237	191	222	260	311	282	271	278	273	
Tr. H.P.		146	204	143	136	161	254	204	219	240	250	163	232
P. H.P. ÷ I.H.P.		0·717	0·757	0·675	0·560	0·610	0·649	0·863	0·758	0·717	0·725	0·690	
Tr. H.P. ÷ I.H.P.		0·417	0·512	0·408	0·400	0·443	0·633	0·567	0·588	0·636	0·645	0·410	0·622

On this occasion a dynamometer of some kind was applied to the "Rattler's" screw shaft, and by it the average thrust amounted to 8722 lbs. Multiplying this by the speed in feet per minute and dividing by 33,000, it will be seen that the power exerted by the screw was 247·8 H.P.; dividing this by the I.H.P. 334·6, the efficiency is shown to be 0·74, which is a very high one; her speed coefficient, however, was 235 against the 245 of her opponent.

The *second trial* was made under steam and sail, the sea being smooth and the wind fair and moderate; the course was from the Yarmouth Roads direct north to a point distant thirty-four nautical miles. The "Alecto" arrived $13\frac{1}{2}$ minutes behind the "Rattler"; her mean speed was therefore 11·2 knots as against 11·9 knots of the "Rattler."

A *third trial* was made, again under steam alone, but this time against a strong wind and heavy sea, on the same northerly course. The distance run on it was, however, sixty miles, which the "Rattler" did in thirty minutes less than the "Alecto," her mean speed being 7·5 knots against that of 7 knots of the paddle vessel. The weather and force of the wind varied considerably throughout the day and affected the two vessels somewhat differently, as might have been supposed; it is also important to note that in the early part of the day the engines of the "Rattler" were developing 364 H.P., while those of the "Alecto" were only able to exert 250 H.P.; but the mean slip of the screw was as much as 42·4 per cent., while the thrust of the screw varied from 5100 lbs. to 12,740 lbs.

Following this run the vessels were tried running before the wind, which still continued to be fairly strong, although the weather had moderated; the speed of the "Rattler" then rose to 10 knots with a slip of only 11·2 per cent., and the power developed by the engines was 369 H.P. On this occasion the thrust of the propeller of the "Rattler" was measured and found to be 9440 lbs., which gave the propeller an H.P. of 290 and consequently an efficiency of 0·78.

The Fourth Trial.—The ships were set on a course with the wind astern, and under sail only; the paddle boards of the "Alecto" were removed beforehand, and the propeller of the "Rattler" was set vertical; and after nearly two hours' running the "Alecto" was more than a mile behind the "Rattler," whose speed had been from 5 to $6\frac{1}{2}$ knots.

The Fifth Trial.—This was again a test under sail only, but made "on a wind" and with all plain sails set; the speed was about $3\frac{1}{4}$

knots, and after five hours' running there was not much difference in the distance run by the two ships, but the "Rattler" had edged up to the windward of the "Alecto."

The sixth trial was also made under plain sail with the wind abeam, the sea being smooth and the breeze freshening, so much so that the "Rattler" attained eventually a speed of 10 knots, and throughout the run of four hours made a mean speed of eight knots and beat the "Alecto" by thirty-eight minutes.

The seventh trial was to test the ability of the "Rattler," under steam only, in towing the "Alecto" with her paddle boards removed, and by the log the speed was found to be about seven knots with I.H.P. of 352, the mean thrust indicated by the dynamometer being 10,178 lbs., giving a propeller H.P. of 223 and an efficiency of 0·638.

The eighth trial was to prove the ability of the "Alecto" to tow the "Rattler," which she did at a speed of little under six knots, but with an I.H.P. less than that developed by the "Rattler" when engaged in the same work.

The ninth trial was perhaps the most interesting to the ordinary observer, and created the greatest sensation, so much so that it has been remembered and quoted time after time since as proving conclusively the superiority of the screw. The two ships were secured stern to stern and the engines of the "Alecto" were started first, and continued running until the "Rattler" was being towed stern first at the rate of about two knots an hour. The engines of the "Rattler" were then started and in five minutes her sternward movement ceased, and from that time onwards the "Rattler" towed the "Alecto" stern first against the full power her engines were capable of developing, and finally attained a speed of 2·8 knots per hour.

Now although, at the time, this last trial was deemed to be a very crucial one, and its result to have been so overpoweringly in favour of the "Rattler" as to settle the question of the superiority of the screw, an examination into the real facts of the case tends to considerably modify such views, while, of course, not reversing the verdict. In no case during the set of trials did the engines of the "Alecto" develop the same power as those of the "Rattler," and unfortunately no attempt appears to have been made to equalise matters in this respect, as might have been done by altering the valve setting or reducing the float area. The propeller of the "Rattler" had been cut down in length previous to these trials, so that its surface was not so great as to prevent a rapid rate of revolution even when moving sternward; and consequently, when opposing the "Alecto" her engines indicated 300 H.P. with a mean thrust of 10,505 lbs. and propeller H.P. of 90½, while the "Alecto's" engines indicated only 141 H.P.

Some further trials were made, of which perhaps the most interesting were those we should call now "progressive trials," wherein the ship was tried at varying reduced speeds; these showed the efficiency to be 0·74 at eight knots and 0·70 at six knots. In a further trial between these two ships, which lasted for seven hours, they were driven hard against a strong wind and a heavy sea, the "Alecto" beat the "Rattler" by half a mile, the mean speed of the screw vessel being only 4·2 knots, but it was in consequence of shortness of steam from having to batten down her engine and boiler rooms to keep the heavy seas from flooding them.

TABLE XXII.—ANALYSIS OF TRIALS OF H.M.S. "RATTLER,"
MADE IN 1843 TO 1845.

	Measured Mile.	Measured Mile.	Nore to Yarmouth.	Off Yarmouth.	When Towing H.M.S. "Alecto."	Progressive	
						Trial No. 1.	Trial No. 2.
Displacement tons	868	1140	1018	1018	1018	1018	1018
Wetted skin sq. ft.	6685	7477	7045	7045	7045	7045	7045
Speed knots	10·074	9·639	9·20	10·00	7·00	8·00	6·00
Indicated horse-power	428	437	335	369	301	205	127
Revolutions per minute	104·0	107·9	94·2	103·6	96·7	84	68
Slip per cent. (apparent)	10·75	17·77	10·2	11·2	33·6	12·2	18·7
Displacement⅔ × speed³ ÷ I.H.P.	217	224	235	277	..	252	172
Thrust by dynamometer . lbs.	..	..	8,722	9437	10,278	6213	4802
" indicated ,	12,346	12,139	10,656	10,540	10,910	7300	5743
" calculated ,,	9364	10,084	7666	9287	8147	6098	3783
Tr. resistance of ship	8996	8681	7453	8806	..	4297	3170
P.H.P. (from calculated thrust)	290	298	216	285	176	149	69·8
Dyn. H.P.	..	..	248	290	223	153	88·4
Tr. H.P.	278	257	210	271	..	106	58·5
P.H.P.÷I.H.P.	0·678	0·682	0·645	0·772	0·585	0·727	0·551
Dyn. H.P.÷I.H.P.	..	..	0·740	0·786	0·740	0·746	0·696
Efficiency Tr. H.P.÷I.H.P.	0·649	0·588	0·621	0·734	..	0·517	0·461

CHAPTER XVI.

TRIALS OF H.M.S. "DWARF" AND OTHER SHIPS, MADE FROM TIME TO TIME BY THE BRITISH ADMIRALTY.

In 1845 the Admiralty had made a series of trials on H.M.S. "Dwarf," a small screw steamer built by G. Rennie in 1841, this being one of the earliest following the "Archimedes." She was purchased by the Government, perhaps for the purpose of these experiments, and was 130 feet long, 16·5 feet beam, and 6·4 feet mean draught; the displacement was 131 tons, the immersed midship section 58·4 square feet, the wetted skin 2365 square feet and the prismatic coefficient only 0·6, so that she had very fine lines.

A large number of trials were made, among them being those scheduled in Table XXIII., which are arranged in three groups of four screws. The groups differ only as to pitch, but the members of each group differ as to surface. They might also be arranged in four groups, each differing only as to surface and the members of each as to pitch.

A dynamometer was used to take the thrust of the screw shaft, which was geared to the engines, having two cylinders 40 inches diameter and 32 inches stroke. The comparatively slow rate of revolution of them should have permitted of the indicator diagrams giving practically the real horse-power developed; but the indicator pipes of those days being often long and of small diameter, the error in power must have been often quite 10 per cent. in a general way and possibly as high as 15 per cent. when the rate of revolution was increased on the introduction of the direct-driven screw. In these trials the dynamometer records are manifestly wrong, being quite inconsistent one with another, and, although Bourne tried to correct them, he has not succeeded in making them intelligible. All the screws were 5 feet 8 inches diameter and of the "Common" kind, with two blades. Their pitch varied from 8 feet with a pitch ratio of 1·42 to

13·23 feet with a pitch ratio of 2·33; their surface area ranged from 8·9 to 22·2 square feet—that is, a surface ratio from 0·356 to 0·888, so that both surface and pitch were high.

It will be seen that the highest general efficiency was obtained with the A3 screw, of 8 feet pitch and 13·3 square feet of surface, and was very fairly high. The lowest efficiency was with the propeller of coarsest pitch and greatest surface.

But while A3 was the most efficient screw, the greatest thrust was developed with A1, the largest surface of the finest pitch group, while the highest speed was obtained with the smallest surface of the same group. The dynamometer gave the highest thrust with C2, a screw of very coarse pitch and considerable surface, but according to it the one with the smallest surface of group C gave very nearly the same thrust. On the other hand, this erratic instrument showed a very low thrust with screw A1, which was scarcely what might have been expected.

Admiralty trials with screw steamers subsequent to those of the "Rattler" and "Dwarf" were as follows :—

In 1847-8 H.M.S. "Minx" was employed to test certain screws differing in surface and pitch, and the dynamometer was again employed to take the thrust. In Table XXIV. are the results and an analysis of them, by which it will be seen that the dynamometer was again at fault, registering about double the actual thrust, and it was inconsistent in its readings, giving 3416 lbs. at 8·54 knots with one screw, and 4713 for the same screw at 8·36 knots. But the highest speed, and that is what they really sought in those days, was obtained with the coarser pitch and the surface a trifle less. The enormous amount of slip with screws having a surface ratio of 0·45 makes one doubt the suitability of the ship for such experiments, and altogether suspicious of the figures given.

In 1857 H.M. yacht "Fairy" was utilised to test the respective merits of various inventors' screw propellers. A few of the results are given in Table XXV.

In 1849 H.M.S. "Archer" was fitted with a screw of very extraordinary proportions, for it was 12·5 feet in diameter and only 7·75 feet in pitch, the pitch ratio being thus only 0·620. After being tried with it, the screw was reduced to 11·62 feet diameter, and tried again; afterwards it was again reduced to 10 feet, tried, and finally reduced to 9 feet diameter.

A new screw was then made 9 feet in diameter, and instead of

TABLE XXIII.—ANALYSIS OF THE TRIALS ON H.M.S. "DWARF" WITH VARIOUS SCREWS, 1845.

Description of Screw.	Common Two Blades.				Common Two Blades.				Common Two Blades.			
	8 feet Pitch.				10·22 feet Pitch.				13·25 feet Pitch.			
	5·67 feet Diameter.				5·67 feet Diameter.				5·67 feet Diameter.			
No. of Trial.	A1	A2	A3	A4	B1	B2	B3	B4	C1	C2	C3	C4
Surface of blades sq. ft.	22·2	17·8	13·3	8·9	22·2	17·8	13·3	8·9	22·2	17·8	13·3	8·9
Pitch ratio	1·42	1·42	1·42	1·42	1·82	1·82	1·82	1·82	2·33	2·33	2·33	2·33
Surface ratio	·888	·712	·532	·356	·888	·712	·532	·356	·888	·712	·532	·356
Revolutions per minute	146·2	152·7	155·4	166·0	123·6	123·3	127·1	139·8	106·6	111·8	114·2	125·9
Slip per cent	25·0	25·7	27·1	30·4	29·3	30·8	33·5	36·4	38·7	39·5	40·4	44·7
Speed of ship knots	8·64	8·96	8·94	9·11	8·89	8·74	8·61	9·05	8·52	8·83	8·83	9·03
Indicated horse-power	130·7	151·5	137·0	168·8	143·8	136·3	148·8	154·0	149·4	166·4	161·7	176·9
Indicated thrust lbs.	3700	4095	3637	4200	3721	3535	3290	3527	3500	3845	3333	3538
Calculated thrust	3587	3579	3137	2940	3304	2964	2733	2714	3179	3142	2848	2820
Dynamometer thrust	3049	3347	3176	3396	2947	2144	2404	1875	3031	3004	2632	2893
Bourne.	2567	3069	2733	3136	2755	2543	2500	2643	2621	2755	2643	2620
Tr. resistance of ship lbs.	2394	2555	2546	2651	2517	2431	2367	2625	2323	2488	2488	2637
Tr. horse-power	63·7	70·3	69·9	74·2	68·8	66·1	62·6	72·8	60·8	67·3	67·6	73·4
Power delivered to screw, N.H.P.	113·2	123·3	108·0	135·3	124·1	116·7	128·3	130·0	134·4	149·4	144·2	156·4
Power delivered by screw Pr. H.P.	95·2	96·4	86·1	82·3	90·2	80·1	72·7	75·3	83·2	85·2	77·4	78·6
Pr. H.P. ÷ N.H.P.	·841	·782	·797	·608	·726	·686	·567	·580	·620	·570	·537	·502
Tr. H.P. ÷ N.H.P.	·563	·570	·649	·549	·562	·567	·488	·660	·452	·450	·469	·469
Tr. H.P. ÷ I.H.P.	·488	·464	·510	·440	·478	·489	·421	·473	·408	·406	·417	·415
Displacement ⅔ × Speed³ ÷ I.H.P.	127	126	139	120	125	141	115	129	106	111	129	111

TABLE XXIV.—ANALYSIS OF CERTAIN TRIALS MADE WITH SCREW PROPELLERS IN 1847-8 ON H.M.S. "MINX," 203 TONS, DIMENSIONS $131 \times 22 \cdot 1$ FEET $\times 5 \cdot 2$ FEET DRAUGHT.

Description of Screw.	All Common Screws with Two Blades 4·5 feet Diameter.					
No. of Trial.	No. 1.	No. 2.	No. 3.	No. 4.	No. 5.	No. 6.
Pitch of screw . . ft.	5·0	5·83	5·83	5·83	5·83	5·0
Surface of blades sq. ft.	7·20	4·93	5·97	7·10	7·10	7·20
Pitch ratio . . .	1·110	1·295	1·295	1·295	1·295	1·110
Surface ratio . . .	0·450	0·308	0·373	0·443	0·443	0·450
Revolutions per minute .	248·9	237·8	232·0	218·7	256·9	250·1
Slip per cent. . . .	32·0	41·7	39·7	37·7	38·0	28·0
Speed of ship knots	8·36	7·97	8·04	7·85	9·13	8·54
Indicated horse-power lbs.	188·1	178·3	177·7	168·4	252·1	193·1
Indicated thrust ,,	5000	4240	4335	4350	5600	5110
Dynamometer thrust ,,	4713	4458	4437	4372	4282	3416
Calculated ,, ,,	2808	2302	2089	2360	3170	2836
Tow-rope resistance of ship lbs.	2169	1963	2002	1909	2576	2262
Pr. H.P. .	72·0	56·4	51·5	56·8	88·8	74·5
Tr. H.P.	55·7	48·1	49·4	46·0	72·6	59·2
I.H.P. delivered to screw N.H.P.	126·2	121·6	123·7	119·1	185·9	130·6
Pr. H.P. ÷ screw friction .	82·0	64·0	60·1	65·3	102·6	84·7
Pr. H.P. ÷ N.H.P. (perpendicular efficiency)	0·574	0·462	0·416	0·477	0·473	0·570
Tr. H.P. ÷ I.H.P. (general efficiency)	0·320	0·292	0·300	0·295	0·310	0·331
Displacement $^{2/3} \times$ speed^3 ÷ I.H.P.	128·6	123·4	125·8	126·6	112·3	125·4

TABLE XXV.—ANALYSIS OF THE TRIALS MADE AT STOKES BAY, 1857, OF VARIOUS PATENT SCREW PROPELLERS ON H.M. YACHT "FAIRY," $144 \cdot 7 \times 21 \cdot 13 \times 6 \cdot 1$; 210 TONS DISPLACEMENT; W.S. 3300 SQUARE FEET.

Kind of Screw.	Scott's.	Griffiths.	Common, both Corners Cut Off.	Boomerang.	Lowes.	Fisher.	Griffiths.	Common.	Loosey.
Diam. of screw ft.	6·5	6·5	6·5	5·83	5·57	6·0	6·5	6·5	6·3
Pitch ,, ,,	8·00	9·70	9·82	7·70	14·12	8·00	8·25	7·90	10·00
Length ,, ,,	1·75	1·54	1·50	2·40	1·66	0·76	1·60	1·83	1·71
Pitch ratio .	1·23	1·49	1·51	1·32	2·54	1·33	1·27	1·22	1·59
Revs. per minute .	177·5	178·7	174·0	226·3	187·5	207·5	212·5	205·0	202·5
Slip per cent. .	19·26	26·88	27·21	31·62	55·37	20·2	23·26	17·15	38·84
Speed of ship knots	11·309	12·463	12·26	11·699	11·658	13·033	13·27	13·229	13·216
Indicated horse-power	359	362	349	335	384	410	410	406	416
Indicated thrust lbs.	8343	6872	6737	6345	4820	8180	7730	8210	6785
Resistance of ship lbs.	5280	6237	6171	5643	5610	6996	7260	7220	7211
Tow-rope H.P. .	183·4	239	230	203	201	280	296	293	293
Tr. H.P. ÷ I.H.P. .	0·510	0·661	0·659	0·605	0·523	0·682	0·721	0·722	0·704
Disp. $^{2/3}\times$ speed^3 ÷ I.H.P.	142·1	188·9	186·7	168·8	144·8	186·5	196·9	198·1	196·0

making the pitch, say, 10 feet, they made it only 7·25 feet, with results very disappointing. The most interesting feature in this trial is the amount of slip, which began with 17·68 per cent. *negative* and ended with 7·24 per cent. *positive*; and that with the new screw 13·59 positive slip was experienced. This would seem to point to magnitude of diameter as the cause of negative slip, or it may be that slip ratio is the measure, if not the cause, of this evil, for in later ships where negative slip occurred the pitch ratio was always small, while the diameter of the screw was often by no means excessive.

TABLE XXVI.—ANALYSIS OF TRIALS MADE ON H.M.S. "ARCHER" IN STOKES BAY, 1849, WITH SCREWS OF DIFFERENT DIAMETERS.

No. of Trial	No. 1.	No. 2.	No. 3.	No. 4.	No. 5.
Diameter of screw ft.	12·50	11·62	10·0	9·0	9·0
Pitch ,, ,,	7·75	7·75	7·75	7·75	7·25
Surface of blades . sq. ft.	48·24	41·2	31·8	26·4	28·5
Pitch ratio .	0·620	0·666	0·775	0·860	0·806
Surface ratio	0·392	0·389	0·405	0·418	0·448
Revolutions per minute	71·01	72·99	98·77	110·2	126·17
Slip per cent.	17·68 neg.	10·19 neg.	0·45 pos.	7·24 pos.	13·59 pos.
Speed of ship . knots	6·39	6·151	7·520	7·818	7·800
Indicated horse-power .	199	206	294	347	420
Indicated thrust lbs.	11,940	12,015	12,686	13,400	15,200
Calculated ,, ,,	6640	5395	6670	6050	7670
Tow-rope resistance of ship ,,	3785	3508	5244	5587	5540
Tr. H.P. .	74·2	66·6	119·6	134·1	133·0
Pr. H.P. delivered by screw	130·5	101·8	153·9	145·8	182·4
Pr. H.P. + friction of screw .	165	127·8	186	178·2	198·0
Pr. H.P. ÷ I.H.P.	0·660	0·494	0·524	0·423	0·434
Tr. H.P. ÷ I.H.P. .	0·372	0·326	0·407	0·387	0·317
Disp. ⅔ × speed³ ÷ I.H.P.	151·7	130·1	166·0	159·0	133·9

In 1856 H.M.S. "Flying Fish," after undergoing some crucial trials, was fitted with a false bow, which not only gave her a finer entrance but made her 18 feet longer. Her best speed as originally built was 11·832 knots with 1362 I.H.P. and coefficient 127·3. After alteration, with the same machinery the speed was 12·572 knots with 1303 I.H.P. and coefficient 165·9. A large number of experiments were made with different screws, including those already alluded to with the Mangin screws.

In 1858 H.M.S. "Diadem" and "Doris," sister frigates of 3800 tons, were tried with screws of different diameters, viz. 18 feet and 20 feet. The "Diadem" with 18 feet showed an efficiency of 0·420, while the "Doris," with 20 feet, had an efficiency of 0·365. But the "Doris" with 18 feet and the same pitch as the 20-feet screw showed an efficiency of only 0·350.

ADMIRALTY EXPERIMENTS ON SCREW BLADE SHAPES.

TABLE XXVII.—ANALYSIS OF EXPERIMENTS MADE WITH VARIOUS SCREWS IN 1859 ON H.M.S. "DORIS," 240 FEET LONG, 48 FEET BEAM, 20·5 FEET DRAUGHT OF WATER, DISPLACEMENT 3714 TONS.

Particulars.		Griffiths Screw, Two Blades.	Common Screw with Two Blades.			
			Complete as First Made.	Leading Corner only Cut Away.	Following Corner only Cut Away.	Both Corners Cut Away.
Diameter of screw	ft.	20·1	20·1	20·1	20·1	20·1
Pitch ,,	,,	32·16	31·95	31·95	31·95	31·95
Acting surface of blades	sq. ft.	106·6	105·3	98·0	98·0	90·8
Pitch ratio		1·607	1·597	1·597	1·597	1·597
Surface ratio		0·341	0·331	0·308	0·308	0·286
Revolutions per minute		49·83	48·25	50·00	49·86	51·83
Slip per cent.		24·23	22·15	23·46	24·75	26·39
Speed of ship	knots	11·981	11·826	12·048	11·816	12·012
Indicated horse-power		2822	2784	2880	2846	2916
,, thrust	lbs.	58,150	59,900	60,000	58,900	58,400
Calculated ,,		33,200	40,800	39,500	39,500	38,800
Tow-rope resistance		28,720	27,980	29,040	27,930	28,850
N.H.P.		2174	2157	2230	2200	2240
T.H.P.		1221	1494	1461	1426	1428
Tr. H.P.		1057	1017	1075	1015	1064
T.H.P ÷ N.H.P.		0·562	0·692	0·655	0·648	0·637
Tr. H.P. ÷ I.H.P.		0·375	0·365	0·373	0·356	0·365
Displacement $^{2/3}$ × speed^3 ÷ I.H.P.		145·6	142·2	145·0	135·8	142·3
Augmented resistance, I.H.P.		164	477	386	411	364
,, H.P. ÷ T.H.P.		0·134	0·319	0·264	0·288	0·268

In 1861 H.M.S. "Duncan," 3985 tons, was tried with several different screws, and among them two were three-bladed with a moderate amount of surface; the one with 115 square feet of surface gave an efficiency of 0·526 against 0·502 of the two-bladed screw with 77·6 square feet of surface; that is, the blades were of the same size and shape but slightly different in pitch.

In 1862 H.M.S. "Shannon," 3612 tons, was fitted and tried with various screws differing in number of blades, surface and pitch. The highest efficiency, 0·445, was obtained with a four-bladed Woodcroft screw as against 0·409, that of a common two-bladed screw. The three- and the six-bladed screws did fairly well, their efficiency being higher than that of the two-bladed. The blades were all of one size and shape.

In 1863 H.M.S. "Emerald," 3563 tons, was tried with several screws, including a common two-bladed, a four-bladed, and a six-bladed, but in this case the six-bladed had not such a huge amount of acting surface, having only 103 square feet against 82·8 on the two-bladed. The efficiency of the six-bladed was 0·466, the highest, as against 0·422 of the two-bladed; the four-bladed was again good, being 0·457, and the highest speed, 12·003 knots, was obtained with it, as also with the four-bladed screw of the "Shannon."

TABLE XXVIII.—TRIALS OF H.M.S. "SHANNON," 3612 TONS, IN 1862, WITH SCREWS VARYING CHIEFLY IN THE NUMBER OF BLADES.

	(1) Common Screw.	(2) Mangin's Patent Screws.	(3) Woodcroft's, Pitch Increasing.	(4) Woodcroft's, Pitch Increasing.	(5) Woodcroft's, Pitch Increasing.
Diameter of screw . ft.	18·10	18·00	18·15	18·15	18·15
Pitch ,, ,,	23·75	24·2–26·0	2·75–24·4	22·7–24·7	22·8–24·4
Number of blades . .	Two	Four	Three	Four	Six
Surface blades, T. area sq. ft.	71·2	113·6	86·4	115·2	172·8
Pitch ratio	1·312	1·444	1·345	1·367	1·345
Surface ratio .	0·272	0·440	0·334	0·446	0·670
Revolutions per minute	58·58	53·17	56·08	53·17	49·67
Slip per cent. . .	17·79	17·42	14·89	10·83	6·34
Speed of ship . knots	11·288	11·330	11·492	11·550	11·208
Indicated horse-power .	2057	2033	2055	2023	1956
Displacement $^{2/3}$ × speed^3 ÷ I.H.P.	164·6	168·4	174·6	179·3	169·5
Indicated thrust lbs.	48,750	48,700	50,000	50,800	53,200
Calculated ,, ,,	34,420	31,611	35,930	37,070	39,530
Resistance of ship . ,,	24,225	24,396	24,080	25,346	23,845
I.H.P. delivered to the screw (N.H.P.)	1559	1581	1579	1571	1534
I.H.P. delivered by the screw (T.H.P.)	1193	1100	1267	1315	1361
Resistance H.P., Tr. H.P.	840	848	851	900	821
Efficiency of screw T.H.P. ÷ N.H.P.	0·765	0·696	0·880	0·837	0·887
General efficiency Tr. H.P. ÷ I.H.P.	0·409	0·417	0·414	0·445	0·420
Tr. H.P. ÷ T.H.P. .	0·704	0·771	0·676	0·684	0·603

In 1865 H.M.S. "Pallas," 3660 tons, was found not to be working efficiently. Two of the four propeller blades were removed, thereby reducing the surface to one-half, and the ship tried again, when it was found the speed was 12·824 with 3630 I.H.P. as against 12·56 knots with 3568 I.H.P. on a lighter draught of water; but with a Griffiths screw having two blades a foot less in diameter, the speed was over 13 knots with 3580 I.H.P. See Table XXX.

TABLE XXX.—TRIALS OF H.M.S. "PALLAS," 225 FEET LONG × 50 FEET BEAM, IN STOKES BAY IN 1865. (i.) ORIGINAL FOUR-BLADED SCREW; (ii.) ORIGINAL SCREW WITH TWO BLADES REMOVED; (iii.) NEW GRIFFITHS SCREW, TWO BLADES; (iv.) GRIFFITHS SCREW HALF BOILER POWER.

No. of Trial		No. 1.	No. 2.	No. 3.	No. 4.
Displacement	tons	2758	3357	3661	3661
Diameter of screw	ft.	19·17	19·17	18·0	18·0
Pitch „	ft.	19·2 – 21·2A	20·6 – 22·6A	19·4	19·4
No. of blades		4	2	2	2
Pitch ratio		1·00 – 1·10	1·08 – 1·18	1·08	1·08
Revolutions per minute		64·25	70·37	79·82	63·02
Slip per cent.		4·13 – 5·72N	10·8 – 18·69P	14·53P	8·16P
Speed of ship	knots	12·654	12·824	13·058	11·078
Indicated horse-power		3568	3630	3581	1906
Indicated thrust	lbs.	86,600	75,100	76,300	51,500
Tow-rope resistance of ship	lbs.	29,700	32,636	34,932	25,092
Tr. horse power		1152	1284	1398	853
Tr. H.P. ÷ I.H.P.		0·323	0·354	0·391	0·448
Displacement $^{2/3}$ × speed^3 ÷ I.H.P.		111·7	130·3	147·7	169·4

The most important experiments made by the Admiralty, outside those made in the tank at Haslar and not published, were those with H.M.S. "Iris," in 1878 (see Tables XXIX., XXXI.). This ship and her sister the "Mercury" were to have made a great advance in speed on any previous ship for naval purposes; moreover, they were twin-screws. Apparently without any previous experiments, model or otherwise, Sir James Wright, then Engineer-in-Chief, specified the screw to be 18·5 feet diameter and 18·2 feet mean pitch with four blades, as shown in fig. 62, and a surface of 97·2 square feet each.

These ships were 300 feet long, 46 feet beam and 18·1 feet mean draught of water, the displacement 3290 tons; the immersed mid-section was 700 square feet; as the prismatic coefficient was only 0·548, the lines were very fine indeed. It was intended that the I.H.P. should be about 7000, and with this it was not unreasonable to expect 17·5 knots.

As a matter of fact the full speed first realised was only 16·58 with 7503 I.H.P., and with 5251 I.H.P. the speed was no more than 15·12 knots. Fortunately Froude had already introduced the progressive trial system, and a few extra runs on the measured mile at reduced speeds permitted the plotting of a curve of I.H.P., a curve of slip, and a curve of indicated thrust. The disclosure made thereby caused the removal of two opposite blades from each screw, and on making a fresh set of trials it was found that 15·726 knots could be made now with 4368 I.H.P., and the fearful loss through superfluous surface and diameter with original screws shown most clearly.

Now, instead of re-setting the two pairs of blades so that the mean pitch was 20 feet, and having another set of trials, two new four-bladed screws were ordered.

The new screws were only 16·3 feet diameter and 20 feet pitch, while the surface was reduced from 97·2 square feet to 72 square feet, with four blades oval or leaf-shaped. The new trials showed that the ship could do 18·573 knots with 7714 I.H.P., or the original speed 16·56 knots with only 5108 I.H.P. Someone, however, still apparently

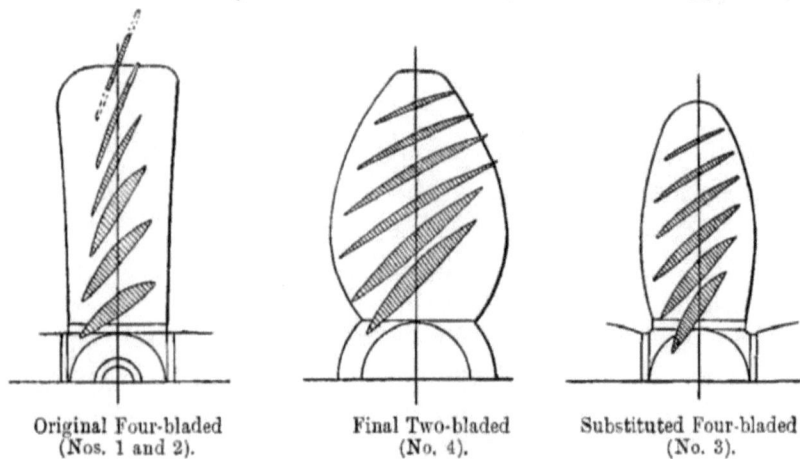

Original Four-bladed (Nos. 1 and 2).   Final Two-bladed (No. 4).   Substituted Four-bladed (No. 3).

FIG. 62.—Screws tried on H.M.S. "Iris."

hankering after the big diameter, now caused another pair of new screws to be made on Griffiths' plan.

These Griffiths screws were with two blades 18·14 feet diameter and 21·3 feet pitch, and an acting surface of only 56 square feet. They were very successful, for the speed now obtained was a trifle higher than with the last screw although the I.H.P. was less.

Now the pity is that the Admiralty did not crop Sir James Wright's screws to 16 feet and try them, or, better still, crop them and increase the pitch to 20 feet; and further, when they were making the Griffiths screws, have put sufficient surface in them so that they would bear cropping to 16 feet.

In order to compare the merits and faults of the four different sets of propellers, the power, revolutions, etc., have been taken from

TRIALS OF H.M.S. "DWARF" AND OTHER SHIPS. 217

TABLE XXXI.—ANALYSIS OF THE TRIALS OF H.M.S. "IRIS," 3290 TONS DISPLACEMENT, 300 FEET × 46·1 FEET × 18·1 FEET DRAUGHT.

No. 1 Screw had four blades, was 18·5 feet diameter and 18·2 feet mean pitch, 97·2 square feet surface each.
,, 2   ,,   two   ,,   18·5   ,,   18·2 ,, ,,   48·6   ,,
,, 3   ,,   four ,,   16·3   ,,   19·96 ,, pitch 72·0 ,,
,, 4   ,,   two   ,,   18·1   ,,   21·3 ,, ,,   56·0   ,,

Speed of Ship	At 9·0 Knots.				At 12·0 Knots.				At 15·0 Knots.				At 18·0 Knots.			
Designation of Propeller	No. 1.	No. 2.	No. 3.	No. 4.	No. 1.	No. 2.	No. 3.	No. 4.	No. 1.	No. 2.	No. 3.	No. 4.	No. 1.	No. 2.	No. 3.	No. 4.
Revolutions per minute	49·1	5·18	46·0	43·0	65·0	67·4	60·1	57·0	81·8	84·2	76·4	73·0		90·0	94·2	90·0
Slip per cent.	1·97N	2·56	1·94	0·43	2·79N	1·14	1·31N	0·16N	2·098	0·60	0·50	2·35		3·40	4·85	
Indicated horse-power	1080	800	820	725	2650	1820	1715	1675	5160	3700	3576	3395		6975	6650	
Indicated thrust lbs.	38,528	28,000	30,240	26,880	63,440	49,290	47,712	43,680	116,480	80,640	77,728	71,680		122,370	114,480	
Calculated	24,970	19,246	20,274	18,527	43,008	32,950	35,410	33,000	67,165	50,292	64,963	52,423		89,923	81,600	
Tow-rope resistance	17,600	17,500	17,500	17,500	31,110	31,110	31,110	31,110	48,400	48,400	48,400	48,400		70,000	70,000	
Augment of resistance	7,470	1,746	2,774	1,027	11,957	1,739	4,299	1,889	18,755	1,692	6,503	4,028		19,923	11,600	
I.H.P. lost in engines	195	208	180	165	290	304	258	240	430	440	373	345		550	500	
I.H.P. delivered to screw N.H.P.	885	592	640	560	2,240	1,516	1,457	1,335	4730	3260	3202	3050		6425	6150	
I.R.P. friction, etc. of screws	175	108	50	51	420	215	100	80	800	425	220	175		370	300	
I.H.P. given out by screws	710	484	590	509	1820	1301	1357	1255	3930	2835	2982	2875		6055	5850	
I.H.P. from calculated thrust T.H.P.	680	532	560	512	1586	1316	1305	1216	3093	2318	2596	2416		4938	4481	
I.H.P. tow-rope resistance Tr. H.P.	483	483	483	483	1147	1147	1147	1147	2229	2229	2229	2229		3870	3870	
Tr. H.P. ÷ I.H.P.	0·997	0·604	0·589	0·649	0·453	0·630	0·669	0·729	0·433	0·603	0·624	0·657		0·555	0·582	
T.H.P. ÷ N.H.P.	0·780	0·914	0·875	0·914	0·799	0·802	0·896	0·911	0·454	0·711	0·803	0·732		0·784	0·729	
Tr. H.P. ÷ T.H.P.	0·700	0·908	0·863	0·943	0·723	0·943	0·879	0·945	0·721	0·961	0·861	0·923		0·784	0·865	
Displacement ⅔ × speed³ ÷ I.H.P.	149	201	196	222	151	210	223	242	145	201	209	219		185	194	

the speed, etc., curves at 18, 15, 12, and 9 knots, instead of simply taking the observed figures, which are all different.

In 1902 four first-class cruisers of the "Drake" class were each fitted with screws differing somewhat in pitch. A speed of nearly 23½ knots was obtained by the "King Alfred" with 31,156 I.H.P., while the "Drake" did 23·05 knots with 30,557 I.H.P. New blades were fitted to the "Drake," having an aggregate surface of 105 square feet as against 76 square feet of the originals. The ship then attained a speed of 24·11 with 31,200 I.H.P., the speed coefficient being 262 as against 234 with the original screws. See Table XXXII.

TABLE XXXII.—ANALYSIS OF THE TRIALS OF FIRST-CLASS CRUISERS WITH SCREWS DIFFERING IN PITCH AND SURFACE, 1903. DIMENSIONS: 500 FEET LONG, 71 FEET BEAM, 26 FEET DRAUGHT WATER; 14,100 TONS DISPLACEMENT.

Name.	H.M.S. "King Alfred."		H.M.S. "Good Hope."		H.M.S. "Drake."		
	Full Power.	Slow Speed.	Full Power.	Slow Speed.	Full Power.	Slow Speed.	Full Power.
Screw, diameter ft.	19·2		19·2		19·2		19·2
,, pitch ,,	23·75		22·8		24·5		23·0
,, number of blades	3		3		3		3
Screw, acting surface, each	76·0		76·0		76·0		105·0
Pitch ratio	1·239		1·19		1·29		1·21
Surface ratio	0·264		0·264		0·264		0·365
Revolutions per minute	120·2	72·6	126·2	77·5	116·0	72·3	122·4
Slip per cent.	16·8	10·5	18·5	8·4	18·0	11·8	13·35
Speed in knots per hour	23·47	15·17	23·05	15·91	23·05	15·43	24·11
Indicated horse-power	31,156	6,743	31,088	7,953	30,557	6,937	31,200
Indicated thrust lbs.	359,400	128,800	356,400	148,560	354,900	129,000	365,630
Calculated ,, ,,	288,473	105,026	304,589	115,214	275,337	108,000	335,954
Tow rope resistance of ship, lbs.	249,603	104,190	240,543	114,609	240,543	107,814	263,193
I.H.P. delivered to screws (N.H.P.)	29,656	6,048	29,463	7,058	29,127	6,227	29,645
I.H.P. given out by screws T.H.P.	20,788	4,891	21,515	5,625	19,498	5,143	24,864
I.H.P. overcoming friction, etc.	2,238	498	2,568	723	2,063	502	3,373
I.H.P. of tow rope (Tr. H.P.)	17,984	4,852	17,038	5,595	17,035	5,100	19,477
Tr. H.P. ÷ I.H.P.	0·576	0·723	0·548	0·703	0·557	0·735	0·624
T.H.P. ÷ N.H.P.	0·701	0·809	0·730	0·797	0·670	0·826	0·840
Tr. H.P. ÷ T.H.P.	0·865	0·992	0·792	0·995	0·874	0·991	0·784
Displacement $^{2/3}$ × speed^3 ÷ I.H.P.	241·5	302·3	230·3	295·7	234·1	309·2	262·0

CHAPTER XVII.

ANALYSIS OF MR CHARLES SELL'S EXPERIMENTS MADE IN 1856 WITH PROPELLERS 6 INCHES DIAMETER VARYING IN PITCH FROM A RATIO OF 0·75 TO 2·25, AND IN SURFACE RATIO FROM 0·316 TO 2·56.

MR CHARLES SELLS was at that time and for many years after the chief of the technical staff of Messrs Maudslay, Sons & Field, and from the first took a strong interest in the screw. He spent much time and labour in these experiments with various screws, and his investigations were made—

(a) At a constant speed of screw of 250 feet per minute (P × R).
(b) At a constant thrust of 2·27 lbs.

Referring to columns A and B (Table XXXIII., No. 1 series), it will be seen that when running so that pitch × revolutions = 250, the total thrust varied from 5·17 lbs. of the finest pitch to 1·26 lbs. of the coarsest screw; but while it required 259 units of power to produce 1 lb. of thrust with the finest pitch, only 171 was necessary with the coarsest.

When running so as to produce the same amount of thrust, the fine pitch screw produced 1 lb. with an expenditure of 170 units, or practically the same as required by the coarsest when running as above. But now the coarsest required for a pound of thrust 230 units of work; but in doing this, the latter screw has a speed of 336 as against the 165 of the fine pitch. Hence work done by the finest pitch is 165 × 1, or 165 foot lbs. work expended to obtain this 170 units.

Comparative efficiency $= 165 \div 170 = 0·97$.

Work done by the coarsest pitch is $336 \times 1 = 336$ foot lbs. work expended to obtain this 230 units.

Comparative efficiency $= 336 \div 230 = 1·46$.

TABLE XXXIII.—No. 1 Series of Mr Charles Sell's Experiments with Six-Inch Model Screws in 1856, Pitch Ratios and Surface Ratios Varying.

Reference Number	Number of Experiments	Propellers, Common.				Working at a Uniform Velocity of 250 Feet per Min.			Working at a Uniform Thrust of 2·27 Pounds.			Relative Efficiencies.	
		Diam.	Pitch.	Pitch Ratio.	No. of Blades.	Surface Ratio.	Thrust of Screw. A	Relative Power Employed. B	Power per Pound of Thrust. C	Velocity of Screw, Feet per Minute. D	Power Employed. E	Power per Pound of Thrust. F	250÷C. G	D÷F. H
		ins.	ins.				lbs.							
I.	3	6·0	4·5	0·75	2	0·316	5·17	1340	259	165	385	170	0·965	0·97
II.	20	6·0	6·0	1·00	2	0·306	3·89	950	245	191	422	186	1·02	1·03
III.	8	6·0	7·5	1·25	2	0·296	2·96	675	228	219	451	198	1·10	1·11
IV.	10	6·0	9·0	1·50	2	0·286	2·27	474	209	250	474	209	1·20	1·20
V.	16	6·0	10·5	1·75	2	0·276	1·96	392	200	269	493	217	1·25	1·24
VI.	11	6·0	12·0	2·00	2	0·266	1·48	268	181	310	510	225	1·38	1·38
VII.	4	6·0	13·5	2·25	2	0·256	1·26	217	171	336	523	230	1·46	1·46

TABLE XXXIII. contd.—No. 2 SERIES. PITCH RATIO SAME THROUGHOUT, BUT SURFACE RATIO AND NUMBER OF BLADES VARYING.

Reference Number.	Number of Experiments.	Propellers, Common.					Working at a Uniform Velocity of 250 Feet per Min.			Working at a Uniform Thrust of 2.27 Pounds.			Relative Efficiencies.	
		Diam. ins.	Pitch. ins.	Pitch Ratio.	No. of Blades.	Surface Ratio.	Thrust of Screw. A lbs.	Relative Power Employed. B	Power per Pound of Thrust. C	Velocity of Screw. Feet per Minute. D	Power Employed. E	Power per Pound of Thrust. F	250÷C. G	D÷F. H
VIII.	12	6.0	9.0	1.50	1	0.143	1.45	284	196	312	554	244	1.275	1.280
IX.	13	6.0	9.0	1.50	2	0.143	1.52	283	186	306	520	229	1.344	1.337
X.	8	6.0	9.0	1.50	3	0.143	1.51	259	173	306	474	209	1.445	1.464
XI.	10	6.0	9.0	1.50	2	0.286	2.27	474	209	250	474	209	1.196	1.196
XII.	9	6.0	9.0	1.50	3	0.286	2.22	435	196	253	450	200	1.275	1.265
XIII.	7	6.0	9.0	1.50	4	0.286	2.22	429	193	253	445	196	1.300	1.291
XIV.	8	6.0	9.0	1.50	6	0.286	2.33	417	180	247	400	176	1.389	1.400

By comparing the work done when running at a uniform speed of 250 feet per minute by 1 lb. of thrust with the power used in producing it; that is, assuming 250 foot lbs. as the amount and dividing it by figures in column C, the comparative efficiency of the different screws can be made. It will be seen on comparing columns G and H that the results are practically the same, and that the most efficient screw is No. VII. and the least is No. I. Strange to say, Mr Sells himself put the values of the screws as propellers in the reverse order, as he took into account only the amount of thrust in each case.

The second series of experiments were made with screws of the same diameter and pitch, but with blades differing only in number; he also made two groups by having Nos. VIII., IX. and X. propellers all of one surface ratio, and Nos. XI., XII., XIII., and XIV. all of one surface ratio, double that of the former.

Judged by the same criterions as before, No. X., having three blades and a surface ratio of only 0·143, is the most efficient screw, and No. XIV., with six blades and a surface ratio of 0·286, follows close on it. By Sell's own criterion No. XIV. was the most efficient, No. X. coming next; he also found No. VIII. to be the least efficient but nearly as good as No. VII., whereas by the criterions here used No. VII. stands as high as No. X., and No. VIII. is better than No. XI., which is a two-bladed screw of 0·286 surface ratio.

It will be seen that at uniform speed the thrust is practically the same for the screws of the same surface ratio (column A) whatever be the number of blades; that it was only the screw with the single blade that differed appreciably from the others of its groups, and the six-bladed which differed from the members of its group.

Comparing the performance of screws which differ only in surface ratio, it will be observed on looking at column A that the thrust of No. IX. is 1·52 lbs. and that of No. XI. is 2·27 lbs.; that is, they are in ratio of 1 to 1·49. Further, the thrust of No. X. is 1·51 lbs., and that of No. XII. is 2·22 lbs.; they are therefore in ratio as 1 to 1·47; that is, the thrusts vary very nearly as the square root of the surfaces.

The blades of No. XIV. must have been very narrow, seeing there were six with a low surface ratio. They were, however, exactly the same in size as those of the screw No. X. Notwithstanding this,

	C1	C2	C3	C4	C5	C6	C7	C8	C9	C10	C11	C12	C13	C14
8th Series. XXXIII. / XXXIV.	1·30 / 1·30	1·30 / 1·31	196 / 206	445 / 468	253 / 268	193 / 191	429 / 378	2·22 / 1·98	4 / 4	·286 / ·286	1·5 / 1·5	9·0 / 9·0	6·0 / 6·0	7 / 2
9th Series. XXXV. / XXXVI.	1·24 / 1·29	1·24 / 1·30	212 / 250	480 / 565	263 / 322	202 / 193	412 / 265	2·04 / 1·37	2 / 2	·238 / ...	1·5 / 1·5	9·0 / 9·0	6·0 / 6·0	14 / 6
Plain blades XXXVII.	1·24	1·24	212	480	263	202	412	2·04	2	·238	1·5	9·0	6·0	14
Blades shrouded on aft side XXXVIII.	1·14	1·13	234	540	267	221	440	1·99	2	·238	1·5	9·0	6·0	9
Blade shrouded on both sides XXXIX.	1·03	1·03	250	570	256	242	530	2·19	2	·238	1·5	9·0	6·0	13
XL. / XLI.	1·52 / 1·43	1·50 / 1·43	225 / 250	510 / 570	341 / 358	166 / 175	202 / 192	1·22 / 1·10	2 / 2	·096 / ·096	1·5 / 1·5	9·0 / 9·0	6·0 / 6·0	7 / 7
XLII. / XLIII.	1·34 / 1·36	1·36 / 1·37	287 / 287	650 / 650	385 / 390	184 / 183	177 / 170	0·96 / 0·93	2 / 2	0·096 / ·096	1·5 / 1·5	9·0 / 9·0	6·0 / 6·0	6 / 5

these two screws proved to be highly efficient, and these results would seem to be confirmed by the experiments made by the Admiralty on H.M.S. "Emerald" in 1863, when general efficiency of 0·466 was shown when running with a six-bladed screw as against 0·457 with a four-bladed and 0·422 with a two-bladed common screw. In previous trials on H.M.S. "Shannon" in 1862 with propellers on Woodcroft's principle, the general efficiency with the six-bladed was 0·420 as against 0·445 with a four-bladed and 0·409 with a two-bladed common screw; but in this case the surface of the six-bladed was very excessive, being 172·8 square feet as against 115·2 of the three- and 71·2 of the two-bladed screws.

The third series of experiments (Table XXXIV.) made by Mr Sells was with two-bladed propellers whose acting surface as well as shape of blade was the same throughout but the pitch differing in each case; in all cases the pitch, as prescribed by Woodcroft, varied from fine at the entering edge to coarse on the following edge of blade. By the same criterions as before, the highest efficiency, 1·38, was with Example XVI., where the variation was from 9 to 13; but Example XV., where it was from 9 to 11, gave an almost equally good result, the pitch ratio being 1·67. Comparing this with the result of the common screw with two blades, No. V., whose efficiency was 1·24 as against the 1·37 of the Woodcroft, the conclusion come to would be in favour of the increasing pitch system. But when screws whose pitch ratio in each case is 2·0 are taken, the Woodcroft shows only 1·36 against 1·38 of the common screw; and going further still to a pitch ratio of 2·25, the Woodcroft is only 1·18 while the common was as high as 1·46.

The fourth series (Table XXXIV.) was with screws of the same diameter, pitch, and with blades of the same size and *shape*, but differing in number, screw No. XXIV. having three blades with a surface ratio of ·429, while screw No. XXVI. had six blades and consequently double the surface ratio.

It will be found on examination that screws Nos. VIII. and XI. are of the same pitch, diameter, shape and area of blade, and differ from the above only in that the first has only one, and the other two blades. Putting these together, then, there is a complete series of screws having blades from one to six in number, which differ only in this way and in the consequent surface ratios.

Taking the thrusts developed by them when running at a constant speed of 250 feet per minute, and dividing them by that of the four-bladed one as the unit, as was done with Froude's and Blechynden's experiments, the following is the result:—

TABLE XXXV.

Number of Blades	1.	2.	3.	4.	6.
Relative thrusts	0·449	0·703	0·907	1·00	1·05
Half square root of number of blades	0·500	0·707	0·866	1·00	1·22

It will also be observed that the efficiency of the screw with six blades and the large surface ratio 0·858 is low; much lower, in fact, than the six-bladed screw with surface ratio of only 0·286, which is what would have been expected.

The fifth series (Table XXXIV.) of experiments consisted of tests of Smith's original screw of one complete convolution with Woodcroft's, also of one convolution. It took the latter 539 units of work to produce a thrust of 2·27 lbs. and Smith's only 427, or practically the same as that of the six-bladed No. XXVII. In fact, Smith's efficiency of 1·28 compares favourably with most of the ordinary screws.

The sixth series (Table XXXIV.) were made with a common screw of two blades and a screw having two circular blades on shanks, all other things being alike; the common screw in this case proved more efficient.

The seventh series (Table XXXIV.) were made with a common and an old Griffiths of the same surface and pitch, etc., each with two blades and with the same surface ratio. Again the common screw more than held its own, and it was better than the Griffiths by 1·38 to 1·36.

Then followed a pair of four-bladed screws, one a common screw and the other a Griffiths. Here the results were practically identical, but very slightly in favour of Griffiths at 250 feet speed.

By comparing XXXI. and XXXIII. it will be seen that the ratio of thrust is 0·685 to 1·0, or the same as found by Froude; doing the same by XXXII. and XXXIV., the ratio is somewhat higher, viz. 0·742 as against 0·707, which is the ratio of the square roots of the surface.

No. XXXV. is a common screw with two blades, and XXXVI. is a screw of the same surface and pitch ratio; the efficiency of the latter was higher than that of the common screw by 1·30 to 1·24.

The next set of experiments is specially interesting to those who suffer from the centrifugal losses, for No. XXXVIII. screw had a flange or shrouding on the aft or acting side, while No. XXXIX. had the shrouding on both sides; the result is seen to be decidedly against these systems, for the efficiency dropped from 1·24 to 1·13 with one set and to 1·03 with two sets of shroudings.

The next four experiments were made with the object of testing the value of the parallel bladed screw, like Taylor's, where the surface was small, giving a surface ratio of ·096.

TABLE XXXVI.—ANALYSIS OF CERTAIN EXPERIMENTS MADE BY MR CHARLES SELLS WITH SCREWS 6 INCHES DIAMETER, 9 INCHES PITCH, PITCH RATIO 1·5, BLADES ALL OF SAME SIZE, EACH HAVING SURFACE RATIO OF 0·143.

Reference Number of Screw.	No. of Experiments.	No. of Blades.	Surface Ratio.	Velocity of 250 Feet.			Thrust of 2·27 lbs.			$\frac{250}{C}$	$\frac{D}{F}$
				Thrust.	Power.	Power per Pound Thrust.	Velocity.	Power.	Power per Pound Thrust.		
				A	B	C	D	E	F		
VIII.	12	1	0·143	1·45	284	196	312	554	244	1·275	1·280
XI.	10	2	0·286	2·27	474	209	250	474	209	1·196	1·196
XXIV.	9	3	0·429	2·93	620	212	220	421	186	1·180	1·180
XXV.	7	4	0·572	3·23	713	221	210	420	186	1·160	1·130
XXVI.	9	6	0·858	3·38	773	229	205	425	187	1·090	1·100

Screws having same shape and size of blades. Common screw.

| Common. XXXI. | 13 | 2 | 0·143 | 1·52 | 275 | 181 | 306 | 500 | 220 | 1·38 | 1·39 |
| XXXIII. | 7 | 4 | 0·286 | 2·22 | 429 | 193 | 253 | 445 | 196 | 1·30 | 1·30 |

Screws having same shape and size of blades. Griffiths.

| Griffiths. XXXII. | 17 | 2 | 0·143 | 1·47 | 270 | 184 | 310 | 575 | 253 | 1·36 | 1·23 |
| XXXIV. | 2 | 4 | 0·286 | 1·98 | 378 | 191 | 268 | 468 | 206 | 1·31 | 1·30 |

CHAPTER XVIII.

EXPERIMENTS MADE BY ISHERWOOD AND OTHERS.

Isherwood's experiments were made in 1869-70 with differing screws on a steam launch of the following dimensions:—

Length on water line	54·4 feet
Breadth, extreme	11·88 ,,
Load draught forward	2·46 ,,
,, ,, aft	3·86 ,,
Immersed mid section	24·98 ,,
Displacement	23·3 tons
Wetted skin total	717 square feet
Rise of floor (angle)	13°-5'

She was propelled by engines having two cylinders 6¾ inches diameter and 8 inches stroke, non-condensing.

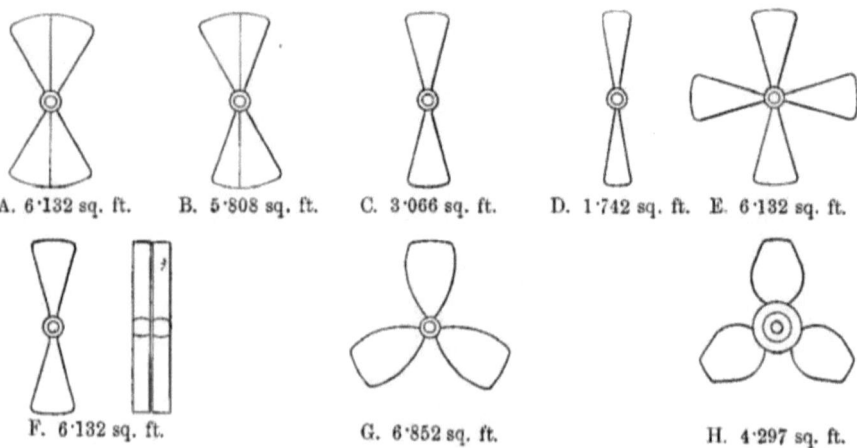

FIG. 63.—Screws used by Isherwood in his Experiments.

Eight different screws were tried. They were made of brass, 52 inches diameter, and their boss was 6 inches diameter.

Propeller A: Common screw with two blades 5·136 feet pitch and 11 inches long; actual surface 6·132 square feet.

Propeller B was a combination of D and C, 8⅝ inches long; actual surface 4·808 square feet.

Propeller C: Common screw with two blades 5·136 feet pitch and 5½ inches long; actual surface 3·066 square feet.

Propeller D: Common screw with two blades 5·136 feet pitch and 3⅛ inches long; actual surface 1·742 square feet.

Propeller E: A combination like A, but the blades at right angles; 6·132 square feet.

Propeller F: The above two screws one in front of the other, 6·132 square feet.

Propeller G: A special screw with pitch 6·5 feet forward and 7·5 feet aft; mean pitch 7·0 feet.

The three blades were curved and 7 inches long at periphery and 6 inches at boss, and had a combined surface of 6·852 square feet.

Propeller H was a Griffiths' three-bladed, with a boss 15 inches diameter and 11 inches long, the surface of blades, 4·297 square feet, and the pitch varying from 6·7 feet forward to 7·33 feet aft; mean, 7·0 feet. In fact, it was made from No. G by cutting it to shape, etc., but otherwise it is hardly fair to call it a Griffiths'.

The following table shows the screws in detail:—

TABLE XXXVII.—PARTICULARS OF THE SCREWS USED BY MR ISHERWOOD IN HIS EXPERIMENTS WITH A STEAM LAUNCH IN U.S. AMERICA, 1869–70.

Designation No.	Description of Screw.	No. of Blades.	Diameter of Screw.	Pitch.	Acting Surface.	Projected Surface.	Pitch Ratio.	Surface Ratio.
			ft.	ft.				
A	Common screw, 11 inches long	2	4·33	5·136	6·132	5·195	1·186	0·409
B	,, ,, (C+D)	2	4·33	5·136	4·808	4·073	1·186	0·321
C	,, ,, 5·5 inches long	2	4·33	5·136	3·066	2·098	1·186	0·204
D	,, ,, 3·125 inches long	2	4·33	5·136	1·742	1·476	1·186	0·116
E	,, ,, right angles	4	4·33	5·136	6·132	5·195	1·186	0·409
F	,, ,, Mangin fashion	4	4·33	5·136	6·132	5·195	1·186	0·409
G	Increasing pitch, curved longitudinally	3	4·33	{6·7 fd. 7·33 aft}	6·852	5·014	1·640	0·457
H	,, ,, Griffiths' shape	3	4·33	{6·7 fd. 7·33 aft}	4·297	2·750	1·640	0·286

And Table XXXVIII. shows the resistance of the boat at speeds varying from 5 knots to 8·5 knots per hour, from which it will be seen that, as in the case of Yarrow's torpedo boat, the law of resistance varying with the squares of velocity does not hold good. Mr Isherwood attributed this to the great variation in trim with variation in speed above 6 knots, consequent on the shortness of the boat. The slip of screws A, E, and F appear to have been identical on all speeds, and varied from 7·82 per cent. at 5 knots to 14·57 at 8·5 knots, which shows that screws of the same surface and shape are not affected by the position of the blades with respect to one another.

TABLE XXXVIII.—RESULTS OF TRIALS AT DIFFERENT SPEEDS.

Speed: Knots per Hour.	Tow Rope Resistance.	Squares of Speed Ratio.	Thrust of the Screw.	Thrust Ratios.	Thrust. I.H.P.
	lbs.		lbs.		
5·0	315·5	1·00	315·4	1·00	4·85
5·5	378·0	1·21	368·8	1·17	6·23
6·0	453·6	1·44	449·9	1·43	8·30
6·5	538·0	1·69	560·6	1·78	11·20
7·0	631·0	1·96	707·0	2·24	15·21
7·5	718·0	2·25	867·1	2·75	19·99
8·0	813·0	2·56	990·8	3·14	24·36
8·5	914·0	2·89	1082·4	3·43	28·28

In comparing the slip of screws A, B, C, and D it is interesting to note that the absolute slip varies in every case as the square root of the acting surface for the same speed of ship.

The slip of No. G, a three-bladed screw with increasing pitch, was greater than that of No. B, in spite of its surface being so much greater when taking the average pitch as the acting one. These experiments would, however, seem to confirm the opinion that such an assumption is not warranted with these screws.

TABLE XXXIX.—SHOWING THE SLIP PER CENT. OF THE PROPELLERS TRIED BY MR ISHERWOOD AT VARYING SPEED.

Ship's Speed: Knots.	Nos. A, E, and F.	No. B.	No. C.	No. D.	No. G.	No. H.
5·0	7·82	8·74	10·43	13·01	9·39	11·60
5·5	8·37	9·35	11·16	13·93	10·58	12·44
6·0	8·87	9·90	11·83	14·76	11·22	13·20
6·5	9·40	10·49	12·54	15·64	11·89	14·01
7·0	10·10	11·26	13·47	16·80	12·77	15·08
7·5	11·56	12·86	15·42	19·83	14·63	17·31
8·0	13·33	14·81	17·78	22·18	16·89	20·06
8·5	14·57	16·15	19·43	24·24	18·48	21·99

TABLE XL.—SHOWING THE PERCENTAGE OF POWER EXPENDED ON PROPULSION BY THE VARIOUS SCREWS TRIED BY MR ISHERWOOD.

Speed of Ship: Knots.	Nos. A, E, and F.	No. B.	No. C.	No. D.	No. G.	No. H.
5·0	79·81	80·71	79·96	78·84	76·78	77·24
8·5	74·05	74·05	71·82	68·43	69·53	68·00

N.B.—During a trial of over nine hours at moorings with screw G the average I.H.P. was 27·22, of which 0·67 was engine friction, leaving,

        Net I.H.P. delivered to shaft, 26·55.

Of this I.H.P.—

Power absorbed by load friction	1·99, or	7·5 per cent.
,, ,, by surface friction of screw	1·46, or	5·5 ,,
,, expended on thrust	23·10, or	87·0 ,,
	26·55	100·0

TABLE XLI.—RESULTS OF TRIALS MADE BY MR ISHERWOOD WITH SCREW B, 4·33 FEET DIAMETER, 5·136 PITCH, AND 4·808 SQUARE FEET OF ACTING SURFACE.

Speed of Ship: Knots per Hour	5·0	8·5
Revolutions per minute	108·2	200·2
Slip of screw per cent.	8·74	16·15
Thrust of screw as taken by dynamometer	315·4	1082·4
Tow-rope resistance of ship	315·5	914·0
Calculated thrust of screw	389·0	1169
Thrust horse-power as deduced from dynamometer	4·85	28·28
Mr Isherwood estimated that of the net thrust horse-power for propulsion per cent.	80·71	74·05
Do. do. friction of screw, ,,	4·06	4·19

TABLE XLII.—RESULTS OF TRIALS MADE AT MOORINGS WITH SCREW G, 4·33 FEET DIAMETER, 7·0 FEET MEAN PITCH, AND HAVING THREE BLADES WITH A SURFACE OF 6·852 SQUARE FEET.

Revolutions per minute	99·0	35·73
Thrust of screw taken by dynamometer lbs.	1093·5	154·50
Gross effective mean pressure on pistons, lbs. per sq. in.	100·0	15·70
Net pressure on pistons	98·0	13·70
Gross effective I.H.P.	28·49	1·60
Net ,, ,,	27·92	1·40
Dynamometer horse-power	24·83	1·26

TABLE XLIII.—RESULTS OF TRIALS MADE WHEN RUNNING FREE FOR 9·18 HOURS WITH SCREW G.

Gross indicated horse-power developed	27·22	27·22	Per cent. of I.H.P.
Power used in working engines and shafting	0·67	...	2·46
,, ,, the friction of load	1·99	...	7·31
,, ,, ,, ,, the screw	1·46	...	5·36
,, expanded in producing thrust	23·10	27·22	84·87

TABLE XLIV.—RESULTS OF TRIALS WITH SCREWS A, E, AND F, SHOWING RELATION BETWEEN INDICATED THRUST AND POWER EXPENDED.

Speeds in Knots	5·0	5·5	6·0	6·5	7·0	7·5	8·0	8·5
Thrust on dynamometer	315·4	368·8	449·9	560·6	707·0	867·1	990·7	1082·4
Tow-rope tension . . lbs.	315·5	373·0	453·6	538·0	631·0	718·0	813·0	914·0
Revolutions per minute	107·13	118·54	130·03	141·69	153·78	167·48	182·30	196·50
Slip per cent.	7·82	8·37	8·87	9·40	10·10	11·56	13·33	14·57
I.H.P. for propulsion	4·847	6·235	8·297	11·200	15·211	19·99	24·361	28·280
,, ,, slip	0·411	0·569	0·808	1·171	1·709	2·613	3·747	4·823
Propulsion and slip	5·258	6·804	9·105	12·371	16·920	22·602	28·108	33·103
Thrust H.P.	5·257	6·805	9·115	12·350	16·910	22·580	28·100	33·100
Calculated thrust . lbs.	375·	..	..	..	..	..	..	1250

TABLE XLV.—SHOWING PRESSURES ON PISTONS DUE TO REACTION OF SCREWS AT VARIOUS SPEEDS. SCREWS 4·33 FEET DIAMETER × 5·136 FEET PITCH.

Screw.	Surface in Feet.		Speed of Ship in Knots per Hour.							
	Helical.	Projected.	8·5	8·0	7·5	7·0	6·5	6·0	5·5	5·0
A, E and F	6·132	5·195	59·076	54·073	47·330	38·584	30·622	24·556	20·129	17·215
B	4·808	4·073	59·074	54·071	47·311	38·588	30·599	24·558	20·129	17·216
C	3·066	2·593	59·085	54·007	47·328	38·595	30·382	24·590	20·091	17·146
D	1·742	1·476	59·200	54·085	47·334	38·583	30·670	24·555	20·134	17·214

Showing Pressures on Pistons due to Reaction of Screws at Various Speeds. Screws 4·33 feet diameter × 7·0 feet mean pitch.

G	6·852	5·014	80·523	73·696	64·501	52·588	41·701	33·467	27·436	23·463
H	5·277	2·750	80·511	73·698	64·504	52·590	41·703	33·468	27·435	23·463

EXPERIMENTS MADE BY ISHERWOOD AND OTHERS.

TABLE XLVI.—DESCRIPTION OF THE PROPELLERS USED BY MR BLECHYNDEN IN THE EXPERIMENTS MADE BY HIM FOR MESSRS R. & W HAWTHORN IN 1882-3.

Distinguishing Mark.	Form of Blades.	Diameter of Screw.	Pitch of Screw.	Number of Blades.	Surface of Blades.	Surface Ratio.	Pitch Ratio.	Remarks.
		ins.	ins.		sq. ins.			
A	Parallelograms	14·5	18·125	4	63·0	0·38	1·250	
B	,,	14·5	18·125	2	31·5	0·19	1·250	Half of above.
C	Modified Griffiths' (merc.)	14·5	15·0	4	63·0	0·38	1·075	
D	,, ,, ,,	14·5	16·0	4	63·0	0·38	1·103	
E	,, ,, ,,	14·5	17·0	4	63·0	0·38	1·172	
F	,, ,, ,,	14·5	18·125	4	63·0	0·38	1·250	Motive weight 2·589.
G	,, ,, ,,	14·5	19·625	4	63·0	0·38	1·353	
H	,, ,, ,,	14·5	21·00	4	63·0	0·38	1·449	
J	,, ,, ,,	14·5	23·20	4	63·0	0·38	1·600	
K	,, ,, ,,	14·5	29·0	4	63·0	0·38	2·00	
L	Proportional to fig. 4.	9·625	12·06	4	27·7	0·38	1·254	Motive weight 1·695.
M	,, ,,	21·75	27·19	4	142·0	0·38	1·250	Motive weight 3·810.
N	,, ,,	21·75	18·13	4	142·0	0·38	0·833	
O	Parallelogram	14·5	18·13	4	92·0	0·556	1·250	
Q	Griffiths' (Naval)	14·5	18·13	4	63·0	0·380	1·250	
P	Fantail, broad tip	14·5	18·13	4	63·0	0·38	1·250	

A very interesting paper communicated by Mr Blechynden to the North-East Coast Institution of Engineers and Shipbuilders in 1886 was practically confined to the enunciation of seven propositions and to the establishment of them by means of the results of the experiments made by Mr Isherwood as already given, and those made by himself for Messrs R. & W. Hawthorn with propellers as per schedule above. Mr Blechynden states that these screws were tried at twenty-three different speeds, the result of which was to indicate that—

1. Thrust $= C_1 \times W$;
2. Thrust $= C_2 \times R^2$;
3. Revs. $= C_3 \times \sqrt{W}$;

R being the revolutions, W the motive weight causing the revolutions, and C_1, C_2, and C_3 are coefficients which are constant for each screw, and any one of which is deduced from the others.

For purposes of comparison they were all reduced to a common thrust of 5·6 lbs., which includes in the indicated thrust an allowance for the effect of fluid friction on the blades. The following are the propositions as enunciated by Mr Blechynden as deduced from the above:—

1. The turning moment in any screw is independent of the quantity of surface or the mode in which it is distributed, friction and edge resistance being excluded.

2. Screws of equal diameter tried under similar conditions have turning moments directly proportional to their pitch ratio for equal thrust.

3. Screws with equal pitch ratio have turning moments proportional to their diameter when indicating equal thrust.

4. Screws tried under similar circumstances have turning moments proportional to their pitches.

5. In any screw, $T \times P = 2\pi \times M =$ thrust $\times$ pitch $= 6.28 \times$ turning moment, which simply means that $T \times P \times R \div 33,000 =$ net H.P.

6. The thrust of any screw working with velocity of advance V and slip S can be approximately determined by the following:—

$$\text{Thrust} = C \times A \times V \times S \times \gamma \div g,$$

A being the disc area of the propeller and γ the density of the fluid; and from experiments made by Mr Blechynden he deduces the value of C as 0·946. The corollary of this is that with constant slip ratio the thrust varies as R^2 or as the square of advance.

7. The effect of surface is the same irrespective of the number of blades into which it is divided, so long as it is similar in distribution.

RENNIE'S EXPERIMENTS

EFFECTS ON THRUST OF IMMERSION OF SCREW

Mr George Rennie in 1856 made an interesting experiment with the object of ascertaining what was the exact benefit to be derived from submerging a screw well below the surface of the water. The records, as reported by him to the Institution of Naval Architects in 1878, while demonstrating that the gain obtained by so doing was great, are not easy to follow as set forth by him in detail. On carefully analysing the figures, however, and dealing with them graphically, the following have been deduced and may be taken as a fairly correct statement of what really took place

The Screw employed was a two-bladed "common" one such as then used in the Navy, 1 foot 9 inches diameter, of true pitch, made of bronze, and highly polished. A tank or caisson was placed and secured on the bank of the Thames close to Rennie's works; the screw shaft projected horizontally through the side and carried the screw at its end about 3·5 feet from it; it was supported close to the screw by a V bracket. Inside the tank the shaft ran in a bearing and was fitted with a belt pulley, by which it was turned at a constant velocity from connections with the factory engine. The thrust was taken on one end of a bell-crank lever; its other end was connected to a balance scale above it, in which weights were placed so as to measure the thrust for the time being. Continuous observations were made from the time the screw was just immersed to that when there was 5 feet of water over the tips. How far the tidal currents affected the screw is not stated, nor can one now say with certainty whether the proximity of the tank was the cause of the erratic records, but is more than probable.

The following may be taken as the real variations in thrust:—

TABLE XLVII.

Distance of Axis below Surface.	Water over Tips.	Thrust.	Distance of Axis below Surface.	Water over Tips.	Thrust.
Feet.	Feet.	Lbs.	Feet.	Feet.	Lbs.
0·875	0·00	46	3·375	2·5	360
1·375	0·50	95	3·875	3·0	372
1·625	0·75	125	4·375	3·5	383
1·875	1·00	160	4·875	4·0	392
2·375	1·50	280	5·375	4·5	400
2·875	2·00	340	5·875	5·0	405

It is remarkable that the thrust should have been more than doubled by an immersion of 6 inches, and shows how very necessary it is under any circumstances to have the blade tips always quite under water. It is perhaps still more remarkable that with a foot of water—that is, the addition of another 6 inches over the tips—there was an increase of thrust of 68 per cent. Most engineers would have thought that the limit had then been reached, and almost useless to go further; but Rennie went on, and found that with 2 feet of water over the tips, equivalent to a little more than one diameter, the thrust had risen to 340 lbs., or more than double that with the 1-foot immersion. After this the thrust continued to increase, but only slowly, so that at 5 feet it was 405 lbs., a pressure it had very nearly reached at 4 feet, for it was then 392 lbs.

Now the acting surface of the screw is not named, but it was probably rather less than one square foot; the pitch also is not known nor are the revolutions stated, but it is obvious that to get a thrust of 400 lbs. with so small a screw they must have been high, perhaps as many as 500 per minute. Now under the circumstances and conditions above named, it is more than likely that this is the first recorded case of cavitation, for no screw properly fed with water could produce such differences in thrust as this one had, nor could a screw which exerts 400 lbs. of thrust fail to produce 50 lbs. fully immersed at the same rate of revolution if fed with water to even quite a moderate amount. The sudden rise in thrust after the immersion reached 1 foot points to cavitation, due to the high rate of revolution and the obstruction of the caisson.

ADMIRALTY EXPERIMENTS
TWIN *VERSUS* SINGLE SCREWS

A review of the performance of the early naval twin screwships will not disclose any convincing superiority as regards speed and efficiency. Table XLVIII. gives the trial results of five sister single-screw ships of the "Nassau" class and five sister twin-screw ships of the "Swallow" class. The older ships were 15 feet longer and 9 inches less beam and of slightly more displacement; in fact, 849 tons as against 716 tons. The twins did a higher speed consequent upon the higher I.H.P., but the Admiralty speed coefficient was only 122 against 143·3 of the "Nassau," and general efficiency was 0·345 against 0·410. It was, however, at half boiler power and cruising speeds that these twins showed to advantage, for with higher speed than the "Nassau," the "Swallow" showed an efficiency of 0·434 and speed coefficient of 154·7 against 0·425 and 147·5 of the "Nassau."

The "Penelope," twin screw of 1867, was of 4368 tons displacement, and did only 12·76 knots with 4703 I.H.P. with a speed coefficient so low as 118, whereas the "Galatea," single-screw frigate of 1860, of 4270 tons, did 13·004 knots with 3516 I.H.P., giving a coefficient of 165. The former ship, however, was handicapped with a double stern. The "Vanguard" performance was so very much superior to the rest of her class, that it is necessary to also take of the same class the "Iron Duke," of 5563 tons. It is seen that her speed was 13·855 with 4789 I.H.P., while the single-screw ship "Triumph," of 6840 tons displacement, required 5114 I.H.P. to do 13·522 knots. But the "Vanguard" herself did 14·94 knots with 5366 I.H.P. and speed coefficient 195 (*vide* Table XLIX.).

ADMIRALTY EXPERIMENTS.

TABLE XLVIII.—TRIALS OF DESPATCH VESSELS AND SMALL CRUISERS. TWIN *VERSUS* SINGLE SCREWS.

Name of Ship.		Single Screws.						Twin Screws.					
		Nassau.	Sylvia.	Myrmidon.	Star.	Coquette.	Average.	Swallow.	Plover.	Bittern.	Seagull.	Ring Dove.	Average.
Length, perpendicular	ft.	185·0	185·0	185·0	180·0	180·0	183·0	170·0	170·0	170·0	170·0	170·0	170·0
Beam, extreme	ft.	28·3	28·3	28·3	28·3	28·3	28·3	29·0	29·0	29·0	29·0	29·0	29·0
Displacement	tons	880	902	775	902	785	848·8	710	787	636	665	718	715·4
Wetted skin		7005	7010	6510	6990	6225	6748	6041	6975	5740	5604	5007	5873·4
Speed, full boiler power		10·585	10·243	10·338	9·757	10·659	10·316	10·847	11·339	11·104	11·224	11·103	11·123
I.H.P.		775	689	782	576	697	703·8	582	977	851	898	945	910·6
Tow-rope H.P.		320	259	275	250	289	284·6	296	339	301	303	330	313·8
Tr. H.P.÷I.H.P. ,,		0·424	0·417	0·352	0·441	0·415	0·409	0·336	0·347	0·354	0·338	0·348	0·344
D$^{2/3}$×S^3÷I.H.P. ,,		144·2	145·6	119·2	148·6	150·8	141·6	121·0	127·2	125·1	129·0	116·0	121·8
Speed, half boiler power		9·16	9·803	8·763	8·066	7·768	8·512	8·718	10·084	9·641	10·019	9·581	9·593
I.H.P.		496	434	404	358	235	385·4	339	536	445	545	463	465·6
Tow-rope H.P.		204	183	168	140	114	161·8	153	240	197	216	199	201
Tr. H.P.÷I.H.P. ,,		0·411	0·422	0·416	0·391	0·485	425	0·461	0·448	0·443	0·400	0·430	·434
D$^{2/3}$·S^3÷I.H.P.		142·7	146·8	140·4	135·1	173·0	147·6	164	161·6	156·5	140·0	152·2	154·6
Engine builders		H. & T.	H. & T.	H. & T.	R. N. & S.	M. R. & S.	..	Rennie	Penn	Avonside	Inglis	M. S. & F.	
Date and place of trial		1866, M.	1866, M.	1867, M.	1866, M.	1862, S. B.	..	1868, P.	1867, M.	1869, P.	1868, P.	1868, S. B.	

TABLE XLIX.—TWIN SCREWS *VERSUS* SINGLE SCREWS. ADMIRALTY TRIALS, 1868 TO 1878.

		Example 1.		Example 2.		Example 3.		Example 4.		Example 5.		Example 6.	
		Iris, T.S.S.	Inconstant, S.S.	Apollo, T.S.S.	Active, S.S.	Captain, T.S.S.	Monarch, S.S.	Vanguard, T.S.S. Half Boilers.	Valiant, S.S.	Iron Duke, T.S.S.	Swiftsure, S.S.	Temeraire, T.S.S.	Sultan, S.S.
Length of ship	ft.	300	337	300	270	320	330	280	280	280	280	285	325
Beam	,,	46	50·3	43·0	42·0	53·2	57·5	54·0	56·0	54·0	55·0	62·0	59·0
Draught-water, mean	,,	18·1	22·6	16·5	18·9	24·8	23·7	21·0	24·0	21·0	24·8	27·0	24·9
Displacement	tons	3290	5328	3400	3057	7672	8070	5563	6123	5563	6537	8571	8714
Wetted skin	sq. ft.	18,600	23,500	15,920	15,900	26,430	27,810	21,178	22,630	21,178	22,920	26,900	28,140
Prismatic coefficient		0·548	0·614	0·583	0·629	0·721	0·709	0·705	0·676	0·705	0·715	0·723	0·711
Speed of ship	knots	16·56	16·51	15·00	14·97	14·24	14·94	12·74	12·65	13·86	13·76	14·65	14·13
Indicated horse-power		5108	7361	2960	4015	5990	7842	2752	3560	4789	4913	7516	8629
Indicated thrust	lbs.	99,000	136,300	58,420	80,000	108,320	154,500	73,000	70,000	102,400	96,700	149,200	165,000
Tr. resistance of ship	,,	63,600	79,900	39,800	44,520	67,065	76,745	42,700	43,260	50,820	52,750	65,100	63,400
Tr. H.P.		3235	4051	1834	2050	2933	3513	1681	1678	2130	2230	2930	2756
Efficiency, Tr. H.P. ÷ I.H.P.		0·633	0·564	0·620	0·510	0·489	0·446	0·611	0·471	0·445	0·454	0·390	0·319
$D^{2/3} \times S^2 \div $ I.H.P.		196·5	186·6	257·6	175·8	187·5	171·0	236·0	190·3	174	185	175	139

It is rather to the merchant ship running without top hamper that one has to look for confirmation that the twin screw ship is more efficient, and this may be seen by referring to Table LIX., where various types of ships are compared with one another.

TABLE L.—ANALYSIS OF EXPERIMENTS MADE BY MR GEO. R. DUNELL WITH AN ORDINARY SCREW AND ONE OF DICKINSON'S SIX-BLADED ON THE S.S. "HERONGATE." 186 FEET LONG × 25·1 FEET BEAM × 19·3 DISPLACEMENT, 1110 TONS, W.S. 7500 SQUARE FEET.

Particulars of Screws, etc.		Ordinary C.I. Mercantile.	Dickinson's C.S. Screws.
Diameter of screw	ft.	10·75	10·5
Pitch ,,	,,	15·0	15·0
Acting surface	sq. ft.	32·0	30·0
Number of blades		4	6
Pitch ratio		1·395	1·430
Surface ratio		0·355	0·346
Revolutions per minute		75·3	73·3
Slip per cent.		20·8	13·0
Speed of ship	knots	8·82	9·43
Indicated horse-power		328·6	389·3
Indicated thrust	lbs.	9600	11,680
Calculated ,,	,,	6938	6734
Resistance of ship	,,	5850	6675
Tr. H.P.		158·5	193·3
D% × S³ ÷ I.H.P.		223	232
Tr. H.P. ÷ I.H.P.		0·482	0·497
T.H.P. ÷ N.H.P.		0·619	0·534
State of bottom		Fresh paint	Foul.
T.H.P.		188·0	194·8

Analysis of some experiments made by Mr Yarrow in 1883 with a 60-ton torpedo boat:—

1. Propelled by her own engines, which were carefully suspended so that the exact thrust could be accurately taken and measured.

2. Towed by another boat at similar speeds after removal of her screw, when the tension on the tow-rope was carefully taken and registered.

The indicated horse-power below is 10 per cent. higher than that registered by the indicators, and is probably nearer the actual gross power developed.

TABLE LI.—YARROW'S EXPERIMENTS.

Speed in Knots	9·0	10·0	11·0	12·0	13·0	14·0	15·0
Indicated horse-power, gross	38·5	49·5	67·1	99·0	143·0	193·6	255·2
Engine friction loss H.P.	9·0	10·1	11·7	13·5	15·4	17·4	19·5
N.H.P., being power delivered to the screws	29·5	39·4	55·4	85·5	127·6	176·2	235·7
Screw friction loss H.P.	1·2	1·4	3·4	12·5	26·6	40·2	65·7
Thrust H.P. by dynamometer	28·3	38·0	52·0	73·0	101·0	136·0	170·0
Augmented resistance H.P.	6·3	8·0	11·0	13·0	18·0	23·0	24·0
Tow-rope horse-power	22·0	30·0	41·0	60·0	83·0	113	146·0
Speed$^3 \div$ constant (288)	25·4	34·5	46·2	60·0	76·3	95·3	117·2
Tr. H.P. $\div$ I.H.P. (general efficiency)	0·571	0·606	0·611	0·606	0·586	0·584	0·573
T.H.P. $\div$ N.H.P. (screw efficiency)	0·960	0·964	0·938	0·854	0·792	0·766	0·720
N.H.P. $\div$ I.H.P. (engine efficiency)	0·766	0·796	0·827	0·864	0·892	0·910	0·923
Augmented R.H.P. $\div$ I.H.P.	0·213	0·203	0·199	0·152	0·141	0·130	0·102
Screw friction $\div$ N.H.P.	0·041	0·036	0·061	0·146	0·208	0·228	10·278

Two-bladed and Four-bladed Screw Propellers.—A series of trials made in 1887 by Mr J. Brucker Andreæ of the Royal Dutch Navy on two ships belonging to their Indian fleet are interesting and instructive, as will be seen by carefully examining the analysis of their results as now given.

The "Ceram" and "Flores" are sister ships by different builders, of the following dimensions:—

Length	152 feet.	Displacement	566 tons.
Beam	25·6 „	Immersed mid section	189 sq. ft.
Mean draught	10·17 „	Wetted skin	4875 „

Engines with three cylinders 20 inches, 29 inches, and 46 inches diameter, 27 inches stroke; working pressure 120 lbs.; three screws tried were of Griffiths' form.

No. 1 screw had four blades and was 9 feet diameter, 13 feet pitch, 30 square feet of surface.

No. 2 screw had two blades and was 9 feet diameter, 13 feet pitch, 15 square feet of surface.

No. 3 screw had two blades and was 9 feet diameter, 11 feet pitch, 17 square feet of surface.

It was noted on the trials that up to 12 knots vibration with the two-bladed screws was not objectionable; over 12 knots it was bad, especially with No. 2. No. 3 screw was tried on the "Flores," which was trimmed more by the stern than was the "Ceram," and hence her screw had better immersion.

EXPERIMENTS MADE BY ISHERWOOD AND OTHERS.

TABLE LII.—ANDREÆ'S EXPERIMENTS.
RESULTS OF TRIALS AT FULL SPEED IN EACH CASE.

Screw.	Revolutions.	Speed.	I.H.P.	Slip per Cent.	Indicated Thrust.	Calculated Thrust.	Tc. H.P.	Tc. H.P. / I.H.P.	$D^{2/3} \times S^3$ / I.H.P.
					Tons.	Tons.			
No. 1	128·6	12·78	804	22·5	7·08	4·76	418	0·520	166
,, 2	132·0	12·23	728	27·7	6·25	3·54	298	0·409	161
,, 3	148·3	12·94	811	24·2	6·91	4·25	378	0·467	171

AT TEN KNOTS WHEN THE E.H.P. WAS ESTIMATED BY MODEL AT 185.

Screw.	Revolutions.	I.H.P.	Slip per Cent.	Indicated Thrust.	$D^{2/3} \times S^3$ ÷ I.H.P.	Estimated Thrust.	Tc. H.P. Estimated.	Tc. H.P. ÷ I.H.P.
						Tons.		
No. 1	91·3	307	16·0	3·82	221	2·39	165	0·537
No. 2	100·0	298	22·5	3·39	205	2·03	140	0·470
No. 3	107·0	332	17·5	3·96	190	2·25	154	0·464

At Speed: Knots	8·70	9·70	10·60	11·35	12·00	11·80
Estimated Net H.P.	118	169	230	298	372	348
Actual I.H.P. with No. 1 Screw	200	300	400	500	600	565
,, ,, ,, 2 ,,	180	275	375	480	625	565
,, ,, of difference	20	25	25	20	25	...
I.H.P. No. 2 ÷ I.H.P. No. 1	0·90	0·92	0·94	0·96	1·04	1·00
E.H.P. ÷ I.H.P. No. 1 screw	0·590	0·563	0·575	0·596	0·620	0·616
E.H.P. ÷ I.H.P. No. 2 screw	0·655	0·617	0·613	0·621	0·595	0·616

At Revolutions	80	90	100	108	116	123	128
Actual I.H.P. of No. 1 screw	200	300	400	500	600	700	800
,, ,, ,, 2 ,,	140	210	300	375	465	545	625
,, ,, of difference	60	90	100	125	135	155	175
I.H.P. No. 1 screw ÷ I.H.P. No. 2 screw	1·430	1·430	1·333	1·333	1·290	1·285	1·280
$\sqrt{30} \div \sqrt{15}$	1·414	1·414	1·414	1·414	1·414	1·414	1·414

EXPERIMENTS MADE BY MR W. G. WALKER IN 1891 WITH THE S.S. "ETHEL," HER DIMENSIONS BEING 55 FEET LONG, 9 FEET BEAM, 3·25 FEET MEAN DRAUGHT, DISPLACEMENT 18·5 TONS, AND WETTED SKIN 733 SQUARE FEET.

The seven screws experimented with had each an acting surface of 2·75 square feet.

No. 1 screw, made up of two halves, each with two blades so set as to form a two-bladed screw.

No. 2 screw, made up of two halves, each with two blades, the blades being parallel as in Mangin style.

No. 3 screw, made up of two halves, each with two blades set with the after one leading the forward.

No. 4 screw, made up of two halves, each with two blades set at right angles.

No. 5 screw, made up of two halves, each with two blades set with the forward one leading by 30°.

No. 6 screw, made up of two halves, each with three blades so set as to form a three-bladed screw.

No. 7 screw, made up of two halves, each with three blades at 60° so set as to form a six-bladed screw.

The speed on each trial was 8·0 miles or 6·94 knots, and the resistance 500 lbs.

TABLE LIII.—WALKER'S EXPERIMENTS.

	No. 1.	No. 2.	No. 3.	No. 4.	No. 5.	No. 6.	No. 7.
Diameter of screws ft.	3·21	3·21	3·21	3·21	3·21	3·21	3·21
Pitch ,, ,,	5·35	5·35	5·35	5·35	5·35	6·00	6·00
No. of blades	2	4	4	4	4	3	6
Pitch ratio	1·667	1·667	1·667	1·667	1·667	1·870	1·870
Revolutions per minute	209·9	212·0	203·0	208·0	218·0	192·2	187·3
Slip per cent.	37·3	37·9	35·2	36·7	39·6	38·7	37·2
Indicated H.P.	30·12	30·45	29·06	29·76	31·34	30·26	29·45
" thrust . lbs.	885	888	844	859	862	866	865
Calculated ,, . . ,,	678	692	605	665	732	629	598
N.H.P. delivered to screws	26·91	27·20	25·95	26·61	28·00	27·32	26·58
Tc. H.P. delivered by screws	14·43	14·72	12·87	14·15	15·58	13·40	12·72
Dispm.⅔ × Speed³ ÷ I.H.P.	77·87	77·03	80·71	78·73	74·80	77·52	79·51
Tc. H.P. ÷ N.H.P.	0·537	0·541	0·492	0·532	0·559	0·491	0·403

H.M.S. "Prince Consort."—In 1863 a series of trials was made with this ship in order to test the efficiency of the different groups of boilers and finding at what speed each set could supply steam to drive the engines. About this time some other ships had a trial at a lower speed than that with half the boilers, but none had so many or any at so low a power as this ship; they thus form a group equivalent to the progressive trials of a modern ship. Table LIV. is an analysis of them, which shows how rapidly the efficiency of the screw rose as the speed fell till 10 knots was reached, when it fell back suddenly; it was never very high, and no doubt accounts for the low general efficiency of this ship at low speeds.

The "Prince Consort" was 273 feet long, 58·4 feet beam and 24·2 feet draught of water, 6430 tons displacement, copper sheathed, and propelled by a single screw 21 feet diameter, and pitch varying from 23·84 feet to 25·7 feet, having four blades and an acting surface of 100 square feet with broad tips; the pitch ratio was therefore 1·22.

It is very obvious from the following analysis that the engine friction and resistances did not with these large horizontal jet condensing engines decrease pro rata with the revolutions. If the power required to drive the ship be deducted from the gross I.H.P. it will be found that at 8 knots it is 549, at 7 knots 485, at 6 knots 450; deducting from these the screw friction, there remain 463, 426, and 420, which looks as if the resistance was practically constant, while the revolutions varied from 34 to 26·5. At 9·0 knots it is only 510 H.P.

242   MARINE PROPELLERS.

TABLE LIV.—ANALYSIS OF THE STEAM TRIALS OF H.M.S. "PRINCE CONSORT," 1863.

Speed: Knots.	Revolutions per Minute.	Gross I.H.P.	E.H.P. Engine Losses.	N.H.P. Power Delivered to Screw.	Sf. H.P. Screw Friction.	Tr. H.P. Remaining to Produce Thrust.	Tc. H.P. Thrust Power as Calculated.	Tr. H.P. Tow-rope Resistance of Ship.	N.H.P. / I.H.P.	Tc. H.P. / Te. H.P.	Tc. H.P. / N H.P.	Tr. H.P. / I.H.P.
13·0	56·0	4150	555	3595	383·0	3212	2007	1865	0·867	0·625	0·558	0·449
12·0	50·8	2850	460	2390	290·0	2100	1600	1458	0·839	0·714	0·628	0·508
11·0	46·2	2120	393	1727	220·0	1507	1307	1125	0·814	0·867	0·757	0·530
10·0	42·0	1650	338	1312	164·0	1148	978	844	0·794	0·852	0·831	0·512
9·0	37·8	1253	292	961	120·0	841	623	615	0·767	0·741	0·648	0·491
8·0	34·0	980	247	733	85·0	648	445	432	0·748	0·687	0·607	0·441
7·0	30·0	775	218	567	58·6	508	305	290	0·732	0·600	0·538	0·374
6·0	26·5	620	170	450	30·0	420	200	170	0·725	0·477	0·444	0·275
5·0	22·8	475	138	337	10·6	326	127	106	0·710	0·398	0·377	0·223

TABLE LV.—ANALYSIS OF PROGRESSIVE TRIALS MADE WITH THE TWIN-SCREW STEAMER "CHELMSFORD," OF 2200 TONS DISPLACEMENT, IN 1893. PRINCIPAL DIMENSIONS, 300 FEET × 34·5 FEET × 13·8 FEET DRAUGHT OF WATER. TWO ENGINES, EACH HAVING CYLINDERS 26–39·5 AND 61 INCHES BY 36 INCH STROKE, EACH PROPELLER 12 FEET DIAMETER, 16 FEET PITCH, AND 39 SQUARE FEET OF ACTING SURFACE.

Speed: Knots.	Gross I.H.P.	Revolutions per Minute.	Engine Losses.	N.H.P. Power Transmitted to the Screws.	Sf. H.P. Screw Losses by Friction etc.	Te. H.P. Power Available for Thrust.	Tc. H.P. Thrust Power of Screws as Calculated.	Tr. H.P. Tow-rope Resistance. H.P.	Tc. H.P. / Te. H.P.	N.H.P. / I.H.P.	Tc. H.P. / N.H.P.	Tr. H.P. / I.H.P.	Sf. H.P. / N.H.P.
18·0	4330	130·0	342	3988	292	3696	3008	2041	0·814	0·921	0·754	0·471	0·073
17·0	3400	122·0	300	3100	240	2860	2495	1700	0·873	0·912	0·805	0·500	0·077
16·0	2770	114·4	265	2505	197	2308	2025	1420	0·878	0·904	0·808	0·513	0·077
15·0	2260	107·0	235	2025	160	1865	1683	1181	0·902	0·896	0·831	0·523	0·079
14·0	1815	100·0	204	1611	131	1480	1378	950	0·931	0·888	0·856	0·524	0·081
13·0	1450	93·0	185	1265	106	1159	1107	775	0·954	0·872	0·875	0·537	0·084
12·0	1128	86·0	163	965	84	881	873	610	0·991	0·856	0·905	0·541	0·086
11·0	887	79·0	142	745	64	681	677	470	0·994	0·840	0·909	0·534	0·086
10·0	677	72·0	122	555	46	509	508	353	1·000	0·818	0·924	0·527	0·085
9·0	489	63·5	105	384	34	350	346	255	0·989	0·786	0·901	0·525	0·084
8·0	354	56·0	89	265	23	242	238	179	0·988	0·749	0·898	0·513	0·084
7·0	254	48·0	75	179	15	164	159	119	0·969	0·701	0·888	0·469	0·084
6·0	180	41·0	61	119	10	109	100	73	0·918	0·661	0·840	0·406	0·084
5·0	128	34·0	50	78	7	71	57	44	0·703	0·609	0·731	0·360	0·084

MARINE PROPELLERS.

TABLE LVI.—EXAMPLES OF BRONZE SCREWS HAVING LOOSE BLADES AS IN H.M.S.

Mark.	H.M.S. A.	H.M.S. B.	H.M.S. C.	H.M.S. D.	H.M.S. E.	H.M.S. F.	H.M.S. G.	H.M.S. H.	H.M.S. J.	H.M.S. K.	H.M.S. L.	H.M.S. M.
Diameter of screw . ft.	16·0	15·0	9·0	19·0	15·75	16·5	14·5	14·5	13·0	10·5	9·25	6·6
Pitch ,, ,,	17·5	16·5	9·6	24·5	20·75	22·0	18·5	16·25	17·0	11·5	10·0	8·6
Number of blades .	2	2	2	3	3	3	3	3	3	3	3	3
Surface ,, sq. ft.	52·6	40·4	14·6	76	85	60	60	44·1	40	24	21·3	12·75
Pitch ratio	1·094	1·100	1·055	1·290	1·317	1·333	1·276	1·121	1·310	1·095	1·081	1·303
Surface ratio	0·262	0·228	0·229	0·269	0·436	0·280	0·364	0·267	0·300	0·277	0·318	0·375
I.H.P. ÷ revolutions	28·8	21·9	3·85	132·0	87·0	52·5	38·0	35·7	33·3	23·0	7·90	5·30
Breadth of blade, max. ins.	64·0	60·0	36·0	54·0	74·0	46·0	49·8	36·0	39·5	29·3	29·3	22·75
Thickness at root ,,	5·63	5·5	3·5	8·90	7·13	7·25	6·5	6·5	6·25	4·9	3·75	2·0
Length of boss ,,	42·0	42·0	18·0	…	44·0	40·6	36·0	37·0	36·0	30·0	25·5	12·7
Diameter of shaft ,,	14·0	12·5	6·75	…	23·5	16·13	14·0	13·9	14·6	12·6	9·0	6·5
Boss made of	A bnze.	A bnze.	A bnze.	A bnze.	Bronze.	A bnze.	A bnze.	A bnze.	A bnze.	A bnze.	A bnze.	Steel
,, weight of . cwt.	51·5	52·5	9·50	…	…	67·5	58·0	40·0	40·0	30·5	15·1	2·8
Blades made of .	A bnze.	A bnze.	A bnze.	…	Z bnze	P bnze.	A bnze.	Z bnze.	A bnze.	Z bnze.	A bnze.	Z bnze.
Blade, weight of each cwt.	36·8	30·0	5·50	…	…	32·30	29·40	21·60	20·80	20·74	6·90	1·41
Total weight of screw ,,	125·0	112·5	20·5	…	…	164·5	146·2	104·8	102·4	62·7	36·0	7·5

Note.—A is Admiralty bronze or gunmetal ; Z is zinc bronze such as supplied by the Manganese Co., Stone & Co., etc. ; P a phosphor bronze.

TABLE LVII.—EXAMPLES OF SOLID CAST-IRON SCREW PROPELLERS AS COMMONLY USED IN THE MERCHANT SERVICE.

Mark		A.	B.	C.	D.	E.	F.	G.	H.	J.	K.	L.	M.
Diameter of screw	ft.	19·5	17·5	16·8	15·5	16·0	13·5	12·0	12·5	11·5	10·8	9·5	8·2
Pitch	,,	26·6	17·5	23·5	20·4	20·5	21·0	17·0	15·75	14·5	15·5	12·75	9·25
Number of blades		4	4	4	4	4	4	4	4	4	4	4	4
Surface	sq. ft.	112	90	92·0	60	77	67	51·2	59	48	42	34·5	25·0
Pitch ratio		1·36	1·00	1·40	1·32	1·28	1·56	1·42	1·26	1·26	1·43	1·34	1·14
Surface ,,		0·341	0·374	0·402	0·318	0·383	0·469	0·453	0·480	0·461	0·461	0·486	0·472
I.H.P. ÷ revolutions		64·70	34·20	36·70	16·00	33·00	23·70	13·70	17·86	11·85	7·11	5·11	2·91
Breadth of blade, max. ins.		46·6	42·0	51·0	35·0	41·5	39·8	32·25	41·0	35·5	31·23	30·5	24·7
Thickness at root		10·0	7·5	9·0	6·5	7·5	7·25	6·25	6·5	6·0	4·8	4·25	3·5
Length of boss		45·0	33·5	42·0	33·0	33·0	33·0	27·0	27·0	27·0	22·0	18·0	15·0
Diameter of shaft		17·75	12·25	13·75	11·0	13·75	12·2	10·25	11·0	10·0	8·50	7·13	6·1
Total weight	cwt.	250·7	116·3	189·0	108·87	113·5	101·0	59·7	69·7	56·3	37·75	27·3	17·1

TABLE LVIII.—EXAMPLES OF SCREW PROPELLERS HAVING BLOOSE LADES AS FOUND IN THE MERCANTILE MARINE, ETC.

Mark.	S.S. N.	S.S. O.	S.S. P.	H.M.S. Q.	H.M.S. R.	S.S. S.	S.S. T.	I.S.Y. V.	T.S.S. W.	T.S.S. X.	T.S.S. Y.	T.S.S. Z.
Diameter of screw . ft.	20·25	23·5	19·0	17·5	19·0	16·0	17·0	16·0	12·0	11·5	10·5	9·0
Pitch ,, ,,	25·0	33·0	23·0	19·5	21·0	23·0	21·5	27·0	16·25	19·0	14·3	11·5
Number of blades .	4	4	4	4	4	4	4	3	3	3	3	3
Surface ,, sq. ft.	120	130	96	86	100	63·6	90	77	42	34	36	26
Pitch ratio .	1·234	1·405	1·210	1·114	1·105	1·437	1·265	1·690	1·354	1·652	1·362	1·278
Surface ratio .	0·373	0·300	0·338	0·359	0·352	0·316	0·400	0·383	0·371	0·327	0·439	0·410
I.H.P. ÷ Revolutions .	62·5	...	41·4	70·8	57·2	25·1	41·1	57·5	20·7	13·5	18·0	8·50
Breadth of blade, max. ins.	...	...	52·5	48·0	52·0	44·8	50·0	37·8	39·5	38·3	44·0	33·0
Thickness at root ,,	8·50	...	5·25	7·25	7·5	6·75	9·25	6·5	5·0	4·25	5·00	3·40
Length of boss . ,,	45	...	43·75	48·0	45	33·0	42·0	45·0	30·0	26	26	22
Diameter of shaft ,,	18·25	...	16·0	17·6	19·0	13·5	15·0	18·0	11·8	10·25	11·75	9·0
Boss made of . .	C steel	C steel	C iron	A bnze.	A bnze.	C iron	C iron	Z bnze.	C steel	C iron	C steel	A bnze
,, weight of . cwt.	...	...	96·0	...	132·7	53·0	74·3	79·55	29·00	26·70	27·25	...
Blades made of . .	Z bnze.	C steel	C steel	Z Bnze.	A bnze.	C iron	C iron	Z bnze.	Z bnze.	Z bnze.	Z bnze.	Z bnze.
Blades, weight of each, cwt.	...	...	48·2	...	48·4	29·3	37·4	45·6	15·00	12·05	11·83	6·5
Total weight of screw ,,	311·5	400·0	289	261·0	326·0	170·2	224	216	74·00	62·75	62·75	...

Note.—C S is cast steel; C I cast iron; A B Admiralty bronze; Z B is bronze as supplied by the Manganese Bronze Co., J. Stone & Co., etc.

T.S.S. Bellona. Full Power.	S.S. Larnica. Full Power.	S.S. Don. Full Power.	S.S. St Ronans. Full Power.	S.S. Cazengo. Full Power.	S.S. Lutterworth.
410′ × 50·7′	460′ × 50′	388′ × 43′	402′ × 43′	340′ × 41′	240′ × 32′
7245	9471	4315	5670	4120	2040
26,800	32,190	19,750	23,600	18,900	11,200
16·5	19·5	19·5	19·0	17·0	13·5
22·0	21·0	26·5	23·0	21·5	17·0
3	4	4	4	4	4
69	118	112	96	90	57
1·334	1·076	1·36	1·21	1·26	1·26
0·323	0·395	0·377	0·339	0·397	0·329
89·5	79·8	72·0	54·5	71·0	91·5
10·90	11·63	10·70	0·81	2·50	13·8
17·33	14·62	16·82	12·28	13·75	13·23
126,500	112,100	80,597	55,300	65,010	33,600
99,422	71,885	61,470	35,060	37,380	27,000
0·786	0·642	0·762	0·634	0·580	0·800
80,400	68,560	55,890	31,050	35,721	19,600
7545	5694	4660	2101	2884	1584
535	453	410	165	212	126
7010	5241	4250	1936	2672	1458
464	407	376	120	282	131
5287	3304	3175	1321	1588	1096
4278	3072	2871	1235	1513	797
1009	232	304	86	75	301
0·928	0·919	0·912	0·922	0·926	0·920
0·754	0·631	0·747	0·682	0·588	0·752
0·567	0·539	0·616	0·588	0·525	0·503
258	246	270	280	245	235

[*To face page* 246.

TABLE LX.—VALUES OF $V^{1\cdot 83}$ AND V^2.

V.	$V^{1\cdot 83}$	V^2	V.	$V^{1\cdot 83}$	V^2	V.	$V^{1\cdot 83}$	V^2	V.	$V^{1\cdot 83}$	V^2
1	1·00	1	21	262·7	441	41	894	1681	61	1842	3721
2	3·56	4	22	286·1	484	42	934	1764	62	1900	3844
3	7·47	9	23	310·3	529	43	976	1849	63	1958	3969
4	12·66	16	24	335·5	576	44	1017	1936	64	2017	4096
5	19·01	25	25	360·7	625	45	1060	2025	65	2077	4225
6	26·50	36	26	388·6	676	46	1103	2116	66	2137	4356
7	35·30	49	27	416·3	729	47	1147	2209	67	2197	4489
8	44·70	64	28	445·0	784	48	1193	2304	68	2258	4624
9	55·70	81	29	474·0	841	49	1238	2401	69	2319	4761
10	67·70	100	30	504·7	900	50	1285	2500	70	2380	4900
11	80·60	121	31	536·0	961	51	1332	2601	71	2442	5041
12	94·20	144	32	568·0	1024	52	1380	2704	72	2504	5184
13	109·0	169	33	601	1089	53	1429	2809	73	2568	5329
14	125·2	196	34	635	1156	54	1480	2916	74	2633	5476
15	142·0	225	35	670	1225	55	1531	3025	75	2700	5625
16	159·7	256	36	704	1296	56	1582	3136	76	2767	5776
17	178·3	289	37	740	1369	57	1634	3249	77	2834	5929
18	198·2	324	38	778	1444	58	1687	3364	78	2901	6084
19	218·8	361	39	816	1521	59	1740	3481	79	2969	6241
20	240·4	400	40	855	1600	60	1795	3600	80	3038	6400

INDEX.

ACTING surface of screw propeller, Seaton's formulæ, 165.
Actual thrust of screw propeller, 116.
Adjusting pitch of screw blades, 28.
Adjustment of pitch, Maudslay, 27.
Admiralty bronze or gun-metal, 195.
    experiments, twin *versus* single screw, 234.
    gun-metal or bronze, 195.
Air resistance, 51.
"Alecto" *versus* "Rattler," 203, 205.
Allen, John, proposals for propulsion, 4.
    system, hydraulic propulsion, 22.
Aluminium bronze, 195.
American Navy triple screws, 144.
Analysis of ship model, Kirk, 43.
    Froude, R. E., 118.
    Froude, Dr W., 121.
    of the performances of sundry steamships on trials, 246.
    of trials, H.M.S. "Archer," 218.
        H.M.S. "Dwarf," 210.
        H.M.S. "Fairy," 211.
        H.M.S. "Minx," 211.
        H.M.S. "Rattler," 205, 207.
    first-class cruisers, 218.
Ancient galleys fitted with paddle-wheels, 3.
Andreæ's trials with two- and four-bladed screw propellers, 238, 239.
Angle of entrance of ship's model, Seaton's rule for, 44.
Apparent slip, 57.
    negative, 59.
"Archer," H.M.S., experiments on, diameter of screw, 209.
    analysis of trials, 212.
"Archimedes," screw steamer, 17, 21.
    s.s., trials of, 202.
Area of paddle floats, 82.
Arms of paddle-wheels, 85.
Atkins, multiplicity of propellers, 33.
Augmented resistance of ship, 50, 109–128.
Axioms relating to screw propellers, 170.

BACON, Roger, observations on ship propellers, 3.
Barber, J., application of turbine to propulsion, 6.

Beaufoy's experiments and formula, 39.
Belfast and Glasgow, steam communication between, 14.
Bell, Henry, built "Comet," p.s., 1811, 10.
Bending moments of propeller blades, curve of, 174.
Bernouilli methods of propulsion, 5.
Bessemer, hydraulic propeller, 97.
    turbine propeller, 23.
Bevis' feathering screw, 159, 161.
Biilington and Newton, bronzes, 199.
Blade area, ratio of thrust to, Sells, etc., 224.
    flange, number of bolts in, 182.
Blades, cast iron, in H.M. service, 200.
    cast steel, 200.
    dovetailed forged bosses, 201.
    forged steel, 200.
    hollow, 190.
    loose—bronze screws, examples of, 244.
    of screws, Griffiths' improvements, 29, 30, 34.
    propeller, effect of surface on thrust, Blechynden, 154.
        Sells, 154.
        Froude, 154.
    effect of number on thrust, 154.
    number of, 151.
    screw, corrugated ribbed, etc., 27.
    various number of, 157.
Blasco de Garay, steamship, 3.
Blechynden, experiments with various screw propellers, 231.
    propeller blades, effect of surface on thrust, 154.
    proposals, 232.
Boss of propellers, Roberts' improvements, 26.
    screw, length of, 168.
Bosses, different experiments with H.M.S. "Conflict," 167.
    elongated, 168.
    paddle-wheel, 84.
Bourne's internal turbo-motor, 28.
Bow screw, Wakefield Pim, 24.
    Howden, J., 33.
Bramah's propeller, 7.
Brandon, s.s., compound engines, 28.

INDEX.   249

Brass, naval, 199.
"Britannia," paddle steamer, 12.
"Britannic," s.s., submersible screw, 149.
Bronze, aluminium, 195.
  manganese, 195, 196.
  phosphor, 195.
  screws, loose blades, examples of, 244.
  Stone's, 195, 197.
Bronzes, Billington and Newton, 199.
Brown, Samuel, chain-driven ferry-boat, 15.
  gas engine, 15.
  special prize for screw propulsion, 1825, 15.
Buckholz, triple screws, 26.
Bull's metal, 199.
Butterly Co., engine builders, 12.

CALLAWAY and Purkiss, hydraulic propulsion, 24.
Carpenter, lifting screw, 21.
  plan for double screws, 26.
Cast iron, 199.
    blades, in H.M. service, 200.
    screw propellers, examples of, 245.
Cast steel blades, 200.
Cavitation, 66.
Centrifugal force, effect on propellers, 183.
  wheel, guide-blades to screws, 24.
"Charkieh," s.s., experiments with three- and six-bladed propellers, 156.
"Charlotte Dundas," experiments with steamer, 10.
"Chelmsford," t.s.s., trials of, analysis, 243.
Chinese use of paddle-wheels, 2.
Church, William, double wheel propellers, 14.
"City of Glasgow," s.s., 25.
Claudius, use of paddle-wheels by, 2.
"Clermont," paddle steamer, 1807, 10.
Coaxial double screws, Captain Smith, 19.
"Comet," H.M.S., paddle steamer, 12, 15.
  loss of s.s. "Comet," 12.
  trials of, 11.
Common screw, 105.
    improved form of, 133.
    surface of, 105.
Compound engine on screw steamer, 28.
"Conflict," H.M.S., experiments with different bosses, 167.
Conoidal screw, Rennie, 20.
Construction of propellers, material used in, 194.
Corrugated screw blades, 27.
Cruisers, "Drake" class, trials with different screws, 218.
  first-class, analysis and trials of, 218.
Cummerow, Charles, screw propeller, etc., 16.
Cunard steamers, four screws, 145.

Curve of bending moments of propeller blades, 174.
  of thrust, Dr W. Froude, 122.
Cylindrical casing to screw, Rigg's proposal, 32.

"DAUNTLESS," H.M.S., experiments with, 146.
  fitted with new stern, 146.
Dawson's service of steamers, London to Gravesend, 1818, 14.
Delta metal, 199.
Depth of water, influence on speed, 51.
Deschamps, diving boat, 28.
Destroyers, resistance, total, of, 53.
  varying, 53.
Details of modern naval screw, 138.
Developed surface, screw propeller, 105.
"Diadem," H.M.S., trials of, 212.
Diameter of bolts of propeller blades, 183.
  of paddle-wheel, 80.
  of propeller boss, 166.
  of screw boss, rule for, 167.
  of screw propeller, 105, 169.
    rule for, 171.
  of screw shaft, rules for, 180.
Differentiation by Froude, Dr W., 123.
Disconnected paddle shafts, Wilkinson, 17.
Diving boat, submersible, Deschamps, 28.
"Doris," H.M.S., trials of, and experiments, 213.
Double propellers, Church's proposals, 14.
  screws, Bennet Woodcroft, 25.
    Carpenter's plan, 26.
    Howden, J., 33.
    Reed, Sir E. J., proposal of, 35.
Droop and Martin, steering steamships, 33.
"Drake" class of cruisers, trials with different screws, 218.
"Duncan," experiments with H.M.S., screws of different number of blades, 213.
  experiments with, 157.
Du Quet, further experiments, 4.
  propeller, 3.
  submarine helix, 5.
"Dwarf," H.M.S., analysis and trials of, 208, 210.
Dynamometer experiments, 117.

EFFECT of centrifugal force on propellers, 183.
  of trim on speed of ship, 52.
Effective horse-power, 58.
  Rankin formula for, 58.
Efficiency, total, of machinery and screw, 116.
  of engine shafting, 116.
  of propeller, 116, 126.
  of propulsion, 49.
  of ship and machinery, 116.
Elongated bosses, 168.

INDEX.

"Emerald," H.M.S., experiments with, 155.
  experiments with different number of blades, 155, 214.
"Enterprise," paddle steamer, Calcutta, 1825, 15.
Ericsson, John, paddle screws, 18, 19.
"Ermack," I.R.S., four screws, 145.
Experiments on immersion of screw, Rennie, 232.
  s.s. "Charkeih," with three- and six-bladed propellers, 156.
  Isherwood's, 154.
  on H.M.S. "Duncan," 157.
  with screws different number blades, 213.
  with different number blades, H.M.S. "Shannon," 213
  "Dauntless," H.M.S., 146.
  "Flying Fish," H.M.S., 212.
  Hirsch screws, 137.
  model screws, Charles Sell's, 219, 220, 221, 222, 223, 224.
  propeller bosses, 167.
  propellers having one, two, three, four, and six blades, each same size, Sell's, 225.
  screws of various surface and number of blades, Walker, 240.
  "Shannon," H.M.S., 157.
  twin screw ships, 142.
  twin *versus* single screw ships, 235, 236.
  various propellers, Blechynden, 231. Isherwood, 226.
  various screws, s.s. "Ethel," Walker, 240.
  Yarrow's, 117, 237.

FAILURE of H.M.S. "Dauntless," 146.
"Fairy," H.M.S., analysis and trials of, 211.
  trials of, 209.
Feathering gear paddle-wheels, 87, 88.
  paddle-wheel, Galloway, patent, 13.
  Morgan, 13.
  Wright's method, 14.
  paddle-wheels, 72.
  screw, Bevis, 159, 161.
  Maudslay, 159, 160.
Ferry-boat, chain-driven, 15.
First-class cruisers, analysis and trials, 218.
Fitch's experiments with steamers, 7, 8.
Fitting loose blades to boss, method of, 182.
Fittings, paddle float, 89.
Flat blades for screws, 162.
Float, mean velocity of, 55.
  pitch of, 82.
Floats, number of, 83.
  propulsion of, 83.
  rules for area, 82.
  steel *versus* wood, 83.

Floats, thickness of, 83.
Flow of water to paddle float, 58.
"Flying Fish," H.M.S., experiments with, 212.
  experiments with Mangin propeller, 133.
Forged bosses, dovetailed blades, 201.
  steel blades, 200.
Formulæ, Seaton's, acting surface of screw propeller, 165.
  diameter propeller bosses, 168.
Four-bladed screws, 163
Four screws on the Cunard steamers, 145.
  in I.R.S. "Ermack," 145.
  of "Lusitania," 145.
  of Mersey ferry steamers, 145.
Fourness and Ashworth paddles, raising, 7.
French Navy, triple screws in, 143.
Frictional resistance of common screw, 111.
  of Griffiths' screw, 112.
  of screw, rule for, 113.
Froude, R. E., analysis of screw trials, 118.
  formula for thrust, 119.
Froude, Dr W., analysis of ship trials, 121.
  curve of thrust, 122.
  differentiation, 123.
  index values for various lengths of ships, 40.
  for various materials, 40, 41.
  propeller blades, effect of surface on thrust, 154.
Fulton, trial experiments, 10.
Fundamental principle of propulsion, 2.

GALLOWAY, Elijah, patent paddle-wheel, 13.
Gas engine, Brown, Samuel, 15.
Gemmell's patent, 19.
Genevois' method of propulsion, 5.
Geometry of screw, 186.
German Navy, triple screws in, 144.
"Great Britain," s.s., 1845, 32.
"Great Eastern," s.s., 1858, 29.
"Great Western," paddle steamer, New York, 1838, 20.
Green, hydraulic life-boat, 97.
Griffiths, improvements in screw blades, 29, 30, 34.
  screw with four blades, 31.
Griffiths' screw patent, 1860, 131.
  frictional resistance of, 112.
  self-adjusting blades, 24.
Guide cylinders to screws, Maudslay, 27.

HADDON, J. C., ribbon blades, 20.
Hale, method of propulsion, 15.
  system hydraulic propulsion, 93.
Hamer, turbine propulsion, 21.
Helix, angle of, 105.
  pitch of, 104.
  surface of, 105.

INDEX.

"Herongate," s.s., trials of, 237.
Hirsch, Herman, improvements in screw propellers, 1866, 32, 36.
  in propeller shafts, 36.
  patent screw, 134.
Hollow screw blades, 190.
  screw shafts, 27.
Hook's windmill compared to propellers, 3.
Horse-power imparted to propellers, 127.
Howden, J., bow screws, 33.
  double screws, 33, 143.
  sheet-metal propeller blades, 35.
  two-screw system, 143.
Hull, Jonathan, patent steamship, 5.
Hunt's steering screws, 20.
Hydraulic life-boat, Green, 97.
  propeller, Bessemer, 97.
    calculations of, 99.
    efficiency of, 101.
  propulsion, 91.
    Callaway and Purkiss, 24.
    Hales system, 93.
    Ramsay system, 92.
    Ruthven, 21, 24, 93.
    Thornycroft system, 95.
"Hydromotor," s.s., 91.

IMMERSION of screw, effect of, 232.
Increasing pitch propeller, proposed by Bourdon and Tredgold, 17.
  Woodcroft's, 17.
Index value, Froude, 40, 41.
Indicated thrust, 114.
Influence on speed of tidal currents, 52.
  of depth of water, 51.
"Iris," H.M.S., propeller blades, effect of surface on thrust, 154.
  trials of, 215, 216, 217.
Iron, cast, 199.
Isherwood's experiments, 154.
  experiments with various propellers, 226.
Italian Navy, triple screws, 143.

JOHNSON'S patent, two pairs of paddle-wheels, 32.
Jouffroy's method of propulsion, 6.

"KINGSTON," paddle steamer, 12.
Kirk, analysis of ship model, 43.
  method of estimating wetted skin, 42.
  proposal for steering steamships, 33.
Krupp, Alfred, forged steel screws, 31.
Kunstadter's patent screws, 147.

LAIRD, screw shaft tunnel, 21.
Lang, Oliver, designed H.M.S. "Comet," 12.
Length of screw propellers, 157.
Lifting screws, Carpenter's proposals, 21.
  Maudslay's plan, 22.
Lignum vitæ, Penn's, 28.
Liquid fuel, 32.

Longitudinal sections of propeller blades, 179.
Loose blades *versus* solid screws, 180.
Lowering screw, Phipps, 1850, 150.
Low's patent screw, 19.
"Lusitania," s.s., four screws, 145.
Lyttleton, William, propeller, 8.

MACINTOSH, reversible blades, 24.
Manby, Aaron, Staffordshire engineer, 12.
Manganese bronze, 195, 196.
Mangin screw, 132.
Marine propulsion, Trevithick patents and proposals, 13.
Marquis of Worcester claimed as inventor, 3.
Materials used in construction of propellers, 194.
Maudslay feathering screw, 159, 160.
  guide cylinders, 27.
  lifting screws, 22.
  method of adjustment of pitch, 27.
  self-adjusting screw blades, 24.
Mean pitch of screw, 125.
  velocity of float, 55.
Mercantile four-bladed screw, 136.
"Mercury," H.M.S., trials of, 215, 216, 217.
Mersey ferry steamers, four screws, 145.
Metal, Bull's, 199.
  Delta, 199.
  Sterro, 199.
"Meteor," H.M.S., triple screws, 1855, 143.
Method of delineating a true screw accurately, 187.
  fitting loose blades to boss, 182.
  forming surface of screw blades, 106, 107, 108.
  founding propeller, 191.
  measuring pitch of propellers, 192.
  propulsion, Brown, Samuel, 15.
    Hale, William, 15.
Miller, Patrick, double-hulled boat, 7.
  experiments with paddle-wheels, 7.
Millington, John, method of propelling vessels, 13.
  steering screw, 14.
"Minx," H.M.S., analysis and trials of, 211.
  trials of, 209.
Modern naval screws, details of, 138.
  practice, number of blades, 158.
Multiplicity of propellers, Atkin's claim, 33.
Multipliers for Seaton formula for wetted skin, 45.
Mumford, estimation of wetted skin, 44.

NAVAL brass, 199.
Napier, David, arrangement of two propellers, 27.
Negative apparent slip of screws, 59.
  slip of screws, examples of, 60, 61.
    explanation of, 62.

INDEX.

Niepce, J. C., hydraulic propulsion, 14.
   internal combustion machine, 14.
Number and position of screws, 140.
   of blades, modern practice, 158.
   of bolts in blade flange, 182.
   of floats, 83.
   of propeller blades, 151.

ONE-PADDLE-WHEELED ship, 76.
Oscillating engines, 1830, Church, 16.
   1827, Maudslay, 16.

PADDLE blades, ribbed or corrugated, 19.
   boat, on Bridgewater Canal, 1793, 8.
   float, flow of water, 56.
      fittings, 89.
      reaction of, 81.
   wheel arms, 85.
      bosses, 84.
      diameter of, 80.
      feathering, 72.
      feathering gear, 87, 88.
      rims, 86.
      ship with wheels, 3, 76.
      ship with one wheel, 76.
      speed of, 56.
      steamers, example of, 79.
      steering of, 84.
   wheels, advantage of, 69.
      two pairs of, Johnson patent for, 32.
"Pallas," H.M.S., trials of, 214, 215.
Pappin's proposals to Prince Rupert, 3.
"Paragon," paddle steamer, 10.
Pattern of propeller blades, 191.
Pearce, perforated blades, 27.
Pearson, Richard, shipbuilder, Yorkshire, 12.
Penn's lignum vitæ, 28.
Perforated blades, Pearce, 27.
Performance of screws, of same surface and pitch, 140.
   of screws varying only in diameter, 139.
   of sundry steamships, 246.
Phipps' submersible screw, 1850, 150.
   lowering screw, 1850, 150.
Phosphor bronze, 195.
Pim, Wakefield, bow screw, 24.
Pitch, increasing screw propeller, 106.
   of float, 82.
   of propellers, method of measuring, 192.
   ratio, influence of, 64.
      screw propellers, 108.
   true, screw propeller, 105.
   variable screw propeller, 105.
   varying screw blade, 190.
Position of quadruple screws, 151.
   of screw, 145.
   of triple screws, 151.
   of twin screws, 150.
Practical method of drawing screw, 189.
Primitive methods of propulsion, 2.
"Prince Consort," H.M.S., trials of, 240.
"Prince of Coburg," paddle steamer, 12.

Principle of propulsion, fundamental, 2.
Progressive trials of t.s.s. "Chelmsford," analysis, 243.
Projected surface screw propeller, 105.
Propeller, abaft rudder, 148.
   blade bolts, number and size of, 183.
   blades, bending moments, rule for, 125.
   dovetailed to boss, 28.
   effect of number on thrust, 154.
      of surface on thrust, 154.
         Blechynden, 154.
         Froude, 154.
         "Iris," H.M.S., 154.
   elliptical section, 31.
   longitudinal sections of, 179.
   number of, 153.
   pattern of, 191.
   sections at root, 190.
      Sells, 154.
   sheet metal, 33, 35.
   thickness at root, rule for, 177.
   thickness of, 173.
      at axis, rule for, 175.
      radially, 177.
      at tip, 177.
   transverse section, 177.
   boss, diameter of, 166.
   experiments with, 167.
   bosses, thickness of metal, 180.
   Bramah, Joseph, 7.
Propellers, efficiency of, 126.
   Hirsch's improvements, 36.
   racing of, 68.
   two, Napier, D., 21.
Proportions of screw blade surface, 164.
Propulsion of floats, 83.
   hydraulic, 91.
      Allen, 92.
      Ruthven, 92.
      Niepce, 14.
   theories of, 54.
Propulsive powers, test of efficiency, 49.
Pumphrey, proposals for twin screws, 16.
Purkiss, Jacob, angle of screw blades, 15.
   patent coaxial screws, 15.
   oblique floats to paddle-wheel, 16.

QUADRUPLE screws, position of, 151.

RACING of propellers, 68.
Radial *versus* feathering wheels, 71.
Ramsay, system of hydraulic propulsion, 92.
Rankin, formula for E.H.P., 58.
Ratio, pitch, influence on 64.
   of thrust to blade area, Sells, 224.
"Rattler," H.M.S., 21-23.
   trials of, 203, 207.
   versus "Alecto," 203, 205.
Reaction of paddle float, 81.
Real slip of screws, 57.
Reed, Sir E. J., double screws, 35.

INDEX.

Rennie, conoidal screw, 20.
  early screw steamers, 28.
  experiments on immersion of screw, 232.
Residual resistance, rule for, 48.
Resistance, augmented, 50, 109, 128.
  frictional, screw propeller, 110.
  of screw, to calculate, 111, 113.
  of destroyers, 53.
  of screw propeller, 109.
  of ships, 39.
  of ship, total, 109.
  of ship, tow-rope, 109, 129.
  of torpedo boat, Yarrow, 237.
  tangential, of ship, 40.
  through air, 51.
  through water of surface of various materials, 48.
Reversible blades to screw propeller, Macintosh, 24.
Ribbon blades, Haddon, patent, 20.
Rigg's proposal, cylindrical casing to screw, 32.
Rims, paddle-wheels, 86.
Roberts' improvements, propeller bosses, 26.
Robertson, oblique floats to paddle-wheels, 16.
"Royal William," paddle steamer, Liverpool to New York, 1838, 20.
  Gravesend, 1838, 20.
Rudder before propeller, 148.
Rudders, double, 21.
Rule for bending moments of propeller blades, 175.
  calculating frictional resistance of screw, 113.
  diameter of screw shaft, 180.
  diameter of propeller bosses, 167.
  screw blades, 164.
  thickness of propeller blades at axis, 175.
  at root, 177.
Rumsay, James, method of propulsion and experiments, 8.
Russian Navy, triple screws, 144.
Ruthven, John, hydraulic propulsion, 21, 24, 93.

"Salamander," H.M.S., 12.
"Savannah," paddle steamer, 1819, 14.
Savory's engines in ship on river Fulda, 4.
  suggestion for steamships, 4.
Schiele turbo-motor, 28.
"Scotia," paddle steamer, built 1862, 31.
Screw blade, surface proportions of, 164.
  with varying pitch, 190.
  blades capable of adjustment, Wain's method, 28.
  methods of forming surface, 106, 107, 108.
  movable in boss, Woodcroft, 22.
  rule for, 164.

Screw blades, shape of, 163.
  boss, length of, 168.
  coaxial, Purkiss, 15.
  common, 105.
Screw propeller, 104.
  actual thrust, 116.
  arrangements for working, Cummerow, 16.
  axioms relating to, 170.
  developed surface, 105.
  diameter of, 105, 168.
    rule for, 171.
  experiment with, Hirsch, 137.
  flat blades for, 162.
  four blades, 162.
    Griffiths, 31.
  frictional resistance, 110.
  geometry of, 186.
  Griffiths' patent, 131.
  Hirsch patent, 1860, 29, 135.
  improvements, Hirsch, 32.
    Thornycroft, 34.
  increasing pitch, 106.
  length of, 157.
  Mangin, 132.
  mean pitch of, 125.
  mercantile, form of, 136.
  method of founding, 191.
  Millington's method of, 13.
  patent, Kunstadter, 147.
  performance of, varying in diameter only, 139.
  pitch of, 105.
  pitch ratio, 108.
  position of, 145.
  practical method of drawing, 189.
  projected surface, 105.
  resistance of, 109.
  surface ratio, 108.
  three blades, 162.
  useful effect of, 128.
  variable pitch of, 105.
  velocity of, 108.
  weight of, 185.
  Woodcroft's increasing pitch proposals, 1832, 17.
propellers, forged steel, Krupp, 31.
  mercantile marine, examples, 245.
  Navy, examples, 244.
  two blades, 158.
  four blades, 162.
  six blades, 237.
  various forms of, 130.
  shaft, diameter of, rules for, 180.
  tunnel, Laird's patent, 21.
  shafts, hollow, 27.
  submersible, 148, 149.
    "Britannic," s.s., 149.
    Phipps, 150.
Screws, same surface and pitch, performance of, 140.
  with increasing pitch, Tredgold, 1827, 17.

Screws, Woodcroft, 1832, 17.
  rectangular blades, Taylor's patent, 19.
Seaton's formulæ for acting surface of screw propeller, 165.
  for calculating thrust, 123, 124.
  for diameter propeller bosses, 167.
  for thickness of screw blades, 173–177.
  method of estimating wetted skin, 44.
  multipliers for formula for wetted skin, 45.
  rule for angle of entrance of ship model, 44.
  frictional resistance of screw blades, 110.
Section of propeller blades, elliptical, 31.
Sections at root of propeller blades, 179, 190.
Self-adjusting blades, Griffiths, 24.
  Maudslay, 24.
Sells, Charles, experiments with model screws, 219, 220, 221, 222, 224.
  experiments with propellers with one, two, three, four, and six blades, 225.
  propeller blades, effect of surface on thrust, 154.
  relation of thrust to blade area, Sells, 224.
Shafts, propeller, improvements, Hirsch, 36.
Shallow-draught screw ships, stern, Thornycroft, 35, 36.
  steamers, Yarrow, 36.
"Shannon," H.M.S., experiments with different number of blades, 213.
  trials with different kinds of screws, 214.
Shape of screw blades, 163.
Sheet-metal propeller blades, 33, 35.
  Howden, 35.
Ships, resistance of, 39.
Shorter, Edward, portable screw propeller of, 1800, 9.
"Sirius," p.s., New York, 20.
Six-bladed screw propeller, trials of, 237.
Skin resistance of ships, 40.
Slip of propellers, apparent, 57.
  real, 57.
  negative, examples of, 60, 61.
  explanation of, 62.
Smith, Captain, coaxial blades, 19.
  Frank Pettit, patent screws, 1836, 17.
Smooth water necessary to high efficiency, 51.
Solid screw v. loose blades, 180.
  cast-iron screws, examples of, 245.
Speed of paddle-wheels, 56.
Steamer service between Glasgow and Belfast, 1818, 14.
Steering of paddle-wheel, 84.
  screws, 147, 149.
    Hunt, 20.
    Millington, 14.
    "Stratheden," s.s., fitted with, 147.

Steering steamships, Droup Martin, 33.
  Kirk's proposals, 33.
Stern, new one fitted on H.M.S. "Dauntless," 146.
  of twin-screw steamer, Dudgeon's proposal, 32.
Stern-wheelers, 76.
Sterro metal, 199.
Stevens, screw propeller turned by rotary engine, 10.
Stone's bronze, 195, 197.
"Stratheden," s.s., fitted with steering screw, 147.
Submersible screw, 148, 149.
Surface ratio proportions of screw propeller, 108.
Symington's experiments with p.s. "Charlotte Dundas," 9.
  patent, 7.
Symons' twin screw, 29.

TAYLOR, James, and Patrick Miller, 7.
  rectangular blades, 19.
"Thames," steamer, made voyage from Glasgow to London, 12.
Theories of propulsion, 54.
Thickness of floats, 83.
  of metal in propeller bosses, 180.
  of propeller blades, 173.
    at root, rule for, 177.
    at tip, 177.
    radially, 177.
Thornycroft's hydraulic propulsion, 95.
  improvements to screw propellers, 34.
  stern for shallow-draught screw ships, 35.
Three-bladed screws, 162.
Three-paddle-wheeled ships, 162.
Thrust, calculations of, 57.
  experiments on, Yarrow, 237.
  Froude's formula, 119.
  Seaton's formula for calculating, 123, 124.
  indicated, 114.
  of screw, 114.
Tidal currents, influence on speed, 52.
Torpedo boat, experiments with, Yarrow, 117.
  resistance of, 237.
Torsion meter, Wimshurst, 1850, 25.
Total resistance of destroyers, 53.
  of ship, 109.
Tow-rope, resistance of ship, 109, 129.
Towing vessels, Chatham Yard, 3.
Transverse sections of propeller blades, 177.
Trevithick, Richard, proposals for marine propulsion, 13.
Trials, analysis and performances of sundry steamships, 246.
  and experiments with H.M.S. "Doris," 215.
  of s.s. "Archimedes," 202.

INDEX.

Trials of H.M.S. "Diadem," 212.
  of s.s. "Herongate," 237.
  of H.M.S. "Prince Consort," 240.
  with different screws, "Drake" class of cruiser, 218.
  H.M.S. "Dwarf," 208.
  H.M.S. "Emerald," different number of blades, 155.
  H.M.S. "Fairy," 209.
  H.M.S. "Iris," 215, 216, 217.
  H.M.S. "Mercury," 215, 216, 217.
  H.M.S. "Minx," 209.
  Mangin screw, on H.M.S. "Flying Fish," 133.
  H.M.S. "Pallas," 214, 215.
  H.M.S. "Rattler," 203, 204.
  H.M.S. "Shannon," different kinds of screws, 214.
  steam, "Waterwitch," H.M.S., 95.
Trim of ship, effect on speed, 52.
Triple screws, American Navy, 144.
  Buckholz, 26.
  French Navy, 143.
  German Navy, 144.
  Italian Navy, 143.
  H.M.S. "Meteor," 143.
  position of, 151.
  Russian Navy, 144.
Turbine propeller, low-pressure, Bessemer patent, 1846, 23.
  propulsion proposed by Hamer, 1843, 21.
Turbo-motor, Schiele, 28.
  Bourne, 29.
Twin screw on one shaft, 147.
  position of, 150.
  ships, experiments with, 142.
  steamer, stern for, 32.
  steamers, early, Rennie, 28.
Twin screws, interlocking, Taylors' patent, 19.
  Symons' arrangements, 29.
  versus single, Admiralty experiments, 234.

Two-bladed screw, 158.
  screws, Howden's system, 143.

USEFUL effect of screw, 128.

VALUES of $V^{1\cdot 83}$ and V^2, 247.
Various forms of screw propellers, 130.
  number of blades, 157.
Velocity of screw propeller, 109.

WALKER, experiments with various screws, s.s. "Ethel," 240.
Water slip, loss by, 126.
H.M.S. "Waterwitch," 93.
  calculations of efficiency, 102.
  steam trials, 95.
Watt, James, suggestion for screw, 6.
Weight of screw propellers, 185.
Wetted skin, Kirk's method, 42.
  Mumford's method, 44.
  numerous examples of, 47.
  Seaton's method, 44.
Wheels, position of, 74.
  without outer rims, 74.
Wilkinson, disconnected paddle-shafts, 17.
Wimshurst torsion meters, 1850, 25.
Woodcroft, Bennet, double screws, 25.
  blades movable in boss, 22.
  patent screw propeller, 17.
Worcester, Marquis of, claimed as inventor, 3.
Wright, Richard, patent feathering paddle, 14.
  two double-cylinder engines, 14.

YARROW, experiment with torpedo boat, 117, 237.
  experiments on thrust and resistance, 237.
  shallow-draught steamers, 36.
"Yorkshireman," paddle steamer, 12.

www.ingramcontent.com/pod-product-compliance
Lightning Source LLC
Chambersburg PA
CBHW050900300426
44111CB00010B/1312